CHEMISTRY AND OUR DAILY EXISTENCE

DR. M.L.BHAT

INDIA • SINGAPORE • MALAYSIA

Notion Press

Old No. 38, New No. 6
McNichols Road, Chetpet
Chennai - 600 031

First Published by Notion Press 2019
Copyright © M.L. Bhat 2019
All Rights Reserved.

ISBN 978-1-64587-089-0

This book has been published with all efforts taken to make the material error-free after the consent of the author. However, the author and the publisher do not assume and hereby disclaim any liability to any party for any loss, damage, or disruption caused by errors or omissions, whether such errors or omissions result from negligence, accident, or any other cause.

No part of this book may be used, reproduced in any manner whatsoever without written permission from the author, except in the case of brief quotations embodied in critical articles and reviews.

FOR MY LOVE

CHAMPA

Contents

Preface 7
Acknowledgement 9

PART I CONCISE BIO-INORGANIC CHEMISTRY 11

    CHAPTER 1 Biologically Important Elements 13

    CHAPTER 2 Toxicity of the Elements 17

    CHAPTER 3 Biochemistry of cell and cellular materials 23

    CHAPTER 4 Biochemistry of Essential Metal Ions 27

    CHAPTER 5 Cytochromes 41

    CHAPTER 6 Oxidative Enzymes 47

    CHAPTER 7 Trace Metal Requirements in Plants and Animals ... 53

    CHAPTER 8 Metalloenzymes 57

    CHAPTER 9 Photosynthesis 69

    CHAPTER 10 Biochemistry of Nonmetals 73

    CHAPTER 11 Problems in Ecosystem 77

PART II HISTORY OF ATOM 89

    CHAPTER 1 Introduction, Democritus' Apple,
    After Long Oblivion 95

    CHAPTER 2 Scientific Basis of the Constitution of Matter,
    Doctrine of Elements 101

    CHAPTER 3 Birth of Spectral Analysis, Discovery of Electron,
    Positive Rays 109

CONTENTS

CHAPTER 4 J.J.Thomson Model, Planetary Model, Rutherford Model, From Classical Mechanics to Quantum Mechanics, Bohr's Model 117

CHAPTER 5 De Broglie Waves, Heisenberg Uncertainty Principle, Schrodinger Wave Equation, Quantum Numbers 125

PART III CHEMISTRY TEACHING AND LEARNING 137

Contents

Preface 7
Acknowledgement 9

PART I CONCISE BIO-INORGANIC CHEMISTRY 11

    CHAPTER 1 Biologically Important Elements 13

    CHAPTER 2 Toxicity of the Elements 17

    CHAPTER 3 Biochemistry of cell and cellular materials 23

    CHAPTER 4 Biochemistry of Essential Metal Ions 27

    CHAPTER 5 Cytochromes 41

    CHAPTER 6 Oxidative Enzymes 47

    CHAPTER 7 Trace Metal Requirements in Plants and Animals 53

    CHAPTER 8 Metalloenzymes 57

    CHAPTER 9 Photosynthesis 69

    CHAPTER 10 Biochemistry of Nonmetals 73

    CHAPTER 11 Problems in Ecosystem 77

PART II HISTORY OF ATOM 89

    CHAPTER 1 Introduction, Democritus' Apple,
    After Long Oblivion 95

    CHAPTER 2 Scientific Basis of the Constitution of Matter,
    Doctrine of Elements 101

    CHAPTER 3 Birth of Spectral Analysis, Discovery of Electron,
    Positive Rays 109

CONTENTS

CHAPTER 4 J.J.Thomson Model, Planetary Model, Rutherford Model, From Classical Mechanics to Quantum Mechanics, Bohr's Model 117

CHAPTER 5 De Broglie Waves, Heisenberg Uncertainty Principle, Schrodinger Wave Equation, Quantum Numbers 125

PART III CHEMISTRY TEACHING AND LEARNING 137

Preface

This book is made up of three parts: Part I, II and III are respectively titled "Concise Bio-Inorganic Chemistry", "History of Atom" and "Chemistry Teaching and Learning".

The contributions of science and technology to our daily existence and vis-à-vis to the environment are indeed too well-known to be listed. The laboratory-oriented Chemistry in all its fields is very much a part and parcel of our daily existence.

Inorganic reactions form the majority of reactions for obtaining energy for living systems. The reactions are of course mediated and made possible by complex biochemical systems. Two interesting aspects of life are: (1) the ability to capture solar energy; (2) the ability to employ catalysts for the controlled release of that energy. Examples of such catalysts are the enzymes which control the synthesis and degradation of biologically important molecules.

Many enzymes depend upon a metal ion for their activity. Metal-containing compounds are also important in the process of chemical and energy transfer, reactions which involve the transport of oxygen to the site of oxidation and various redox reactions resulting from its use. The Part I of this book "The Concise Bio-Inorganic Chemistry" is devoted to this blooming field of bio-inorganic chemistry. This part covers: the biologically important elements: the essential and trace elements and the biochemistry of essential metal ions. Important topics such as cyanide poisoning, toxicity of metals and problems of our ecosystem have been discussed.

Part II of this book "The History of Atom" was written to follow through the whole path of development of the structure of atom. This path is stretched from 95–55 B.C. to 1927 A.D. The history of atom has been developed from the views of ancient Greek philosopher Democritus's Apple (460 B.C) through classical mechanics, models of atom, de Broglie waves, Heisenberg uncertainty Principle to Schrodinger's Wave Equation (1927).

PREFACE

   Part III deals with Chemistry teaching and learning. The four-component system of Student, Curriculum, Society and Teacher (SCST System) has been highlighted. The course content, the syllabi and the quality and quantity of the courses of study have been emphasized. The training of teachers, the method of assessment of students, and various facilities to impart sound education in chemistry are the main content of this part. The instruments, the necessary equipment, the laboratory facilities and the requirements of library have been stressed. The observation, handling and storage of hazardous chemicals need special attention in a chemistry laboratory.

DR. M.L. BHAT

Acknowledgement

Many books and journals of international repute have been consulted to write this book. The list is given in 'References'. I am highly indebted to respected authors James E. Huheey, J. D. Lee, M. Karapetyants. S. Drakin, Russel S. Drago, A. K. Barnard, V. J. Rydnik, Cotton and Wilkinson.

I would like to take this opportunity to thank my great teacher late Dr. C. N. Kachroo and my teacher and research -guides Late Dr. K. P. Dubey of Kashmir University and Dr. M. L. Tobe of University College London. I also consider myself very fortunate to have been associated with some very fine colleagues.

In writing this book, I would like to express my heartiest thanks to my wife and batch mate Champa for her unflinching patience. My son Nitin deserves special thanks for gifting me the laptop. Thanks to Pramila, my daughter-in-law for assisting me in compiling this book. I am highly thankful to my daughter Dr. Nirupama and her husband Dr. Alok for their enthusiasm in going through the manuscript.

DR. M.L. BHAT

PART I
CONCISE BIO-INORGANIC CHEMISTRY

CHAPTER 1

Biologically Important Elements

The number of elements that are known to be biologically important comprises a relatively small fraction of the 115 known elements. Natural abundance limits the availability of the elements for such use. Molybdenum ($Z = 42$) is the heaviest metal, and iodine ($Z = 53$) is the heaviest non-metal of known biological importance. The metals of importance in enzymes are principally those of the first transition series, and the other elements of importance are relatively light: sodium, potassium, magnesium, calcium, carbon, nitrogen, phosphorus, oxygen, chlorine, and of course, hydrogen.

I. Essential and Trace Elements

An **essential element** is that which is absolutely necessary for life processes. It is very difficult to determine the "essentiality" of an element. An *essential* element may be required perhaps in large, perhaps in small quantities. The phenomenon of an essential element becoming toxic at higher than normal concentrations is not rare. Selenium is an *essential* element in mammals yet one of the most vexing problems is the poisoning of livestock from eating plants that concentrate this element.

The *essential* transition elements are:

Fe, Co, Zn, Cu, V, Cr, Mn, Ni, Mo.

Representative essential metals are:

Na, K, Mg, and Ca

Essential non-metals are:

C, N, O, P, S, Cl

All of these elements except Mo are relatively abundant in the earth's crust. When we look for abundant elements that are not essential elements, we find only four—Si, Al, Ti, and Zr—all of which form extremely insoluble oxides at biologically reasonable pH values.

The term "trace element' although widely used is not precisely defined. However, a **trace element** may be broadly defined as one that is required in smaller quantities. For example the trace element selenium is required at a level of no less than 0.4 ppm in the diet of cattle but can be toxic at levels greater than approximately 4 ppm. Molybdenum averages about 1–2 ppm (*parts per million*) in rocks, soils, plants, and marine animals and even lower in land animals. Yet it is an essential trace metal. At the other extreme, iron, which averages about 5% in rocks and soils and 0.02–0.04% in plants and animals, might or might not be considered a "trace" metal. The importance of trace elements is manifold.

The recent development of analytical techniques capable of determining *parts per billion* has opened new vistas for the study of trace elements. Elements not required by animals and plants for survival, are called **nonessential elements**. Nonessential heavy metals such as lead, mercury, arsenic, cadmium, and chromium are usually toxic at much lower levels than trace essential elements.

II. Functions of Elements in Biological Systems

The chemistry of life can ultimately be referred to two chemical processes: 1. The use of radiant solar energy to drive chemical reactions that produce oxygen and reduced organic compounds from carbon dioxide and water, and 2. The oxidation of the products of (1) with the production of carbon dioxide, water, and energy.

An understanding of our fragile environment can begin with recognition of certain elements in the lives of humans and animals and in the soils that support plants. The biogeochemistry of life as regards to essential elements is given in **Table I:**

Table I. The Bio-Geo-Chemistry of Life—Essential Elements

Essential to all animals and plants:	H, C, N, O, Na, Mg, P, S, Cl, K, Ca, **Mn, Fe, Zn, Se**
Essential to several classes of animals and plants:	Si, **V, Co, Mo,** I
Essential to a wide variety of species in one class:	**B, F, Cr, Br**
Essential to one or two species only:	**Li, Al, Ni, Sr, Ba**
Recent work indicates essentiality, but of unknown function:	**Rb, Sn**

Elements in bold type are generally considered to be trace elements

The functions of the elements in biological systems are given in **Table II:**

Table 1I. Functions of Elements in Biological Systems

At.No. Element	Biological functions
1 Hydrogen	Constituent of water and All organic molecules. Molecular hydrogen Metabolized by some bacteria
6 Carbon	Synthesis of all organic molecules and of biogenetic carbonates.
7 Nitrogen	Synthesis of proteins, nucleic acids etc; Steps in the Nitrogen Cycle are important activities of certain microorganisms.
8 Oxygen	Structural element of water and most organic molecules in biological systems; required for respiration by most organisms.
9 Fluorine	Essential element, 2.5 ppm in diet for optimal growth. Strengthens teeth in mammals.
11 Sodium	Important in nerve functioning in animals. Major component of vertebrate blood plasma. Activates some enzymes.
12 Magnesium	Essential to all organisms. Present in all chlorophylls. Has other electrochemical and enzyme-activating enzymes.
15 Phosphorus	Important constituent of DNA, RNA, bones, teeth, some shells, membrane phospholipids, ADP and ATP, and metabolic intermediates.
16 Sulphur	Essential element in most proteins; Important in tertiary structure of proteins; involved in vitamins, fat metabolism, and detoxification processes
17 Chlorine	Essential for higher plants and mammals. NaCl electrolyte; HCl in digestive juices.
19 Potassium	Essential to all organisms with the possible Exception of blue-green algae; major cation of cytoplasm; important in nerve function and cardiac function.

20 Calcium	Essential for all organisms; used in cell walls, bones, and some shells as structural component; important electrochemically and involved in blood clotting.
24 Chromium	Essential, functioning as a glucose tolerance factor It is related to insulin in its biological role and thus to sugar metabolism and diabetes.
25 Manganese	Essential to all organisms; activates Numerous Enzymes; deficiencies in soils lead to infertility in mammals, bone malformation in growing chicks.
26 Iron	Hemoglobin and Myoglobin are iron- containing complexes. Oxygen transport in vertebrates.
27 Cobalt	Essential for many organisms including mammals. activates a number of enzymes; Vitamin B_{12}.
28 Nickel	Essential trace element. Chicks and rats raised on deficient diet show impaired liver function and morphology. Stabilizes coiled ribosomes.
29 Copper	Essential to all organisms; Constituent of Redox enzymes and hemocyanin.
30 Zinc	essential to all organisms; used in enzymes; stabilizes coiled ribosomes. Plays a role in sexual maturation and reproduction.
42 Molybdenum	Essential to all organisms with the Possible exception of green algae; used in Enzymes concerned with nitrogen fixation and nitrate reduction.
53 Iodine	Essential to many organisms; Thyroxin important in Metabolism and growth regulation.

CHAPTER 2

Toxicity of the Elements

The problem of toxicity is difficult to quantify. There are many synergistic effects between various components of biological systems that it is almost impossible to define the limits of beneficial and detrimental concentrations. There is also endless variation among organisms. Truly, "one man's meat is another man's poison." No common element is toxic at levels normally encountered, though of course almost anything can be harmful at too high levels. When we consider the elements that are currently causing problems in the environment, we find that they are all extremely rare in their crustal abundances: As (0.024), Pb (0.08), Cd (0.0018), and Hg (4×10^{-5}). The conclusion is inescapable: *Life evolved utilizing those elements that were abundant and available to it and became dependent upon them*. Those elements that were rare were not used by living systems because they were not available; neither did these systems evolve mechanisms to cope with them. Many elements that are *essential* when occurring at ambient concentrations are *toxic* at higher concentrations (and, of course, cause deficiency symptoms at lower concentrations). Interesting examples are copper, selenium, and even sodium. NaCl and our blood has been described as a sample of primeval seas. Yet too high concentrations of NaCl are toxic through simple hypertonicity, i.e; osmotic dehydration. Selenium is a problem when it is either too rare or too abundant in the environment. Livestock grown on selenium-deficient pasture suffer from "white muscle disease"; when grazing plants that concentrate selenium from the soil, they suffer central nervous system toxinosis. Copper is essential to many of the redox enzymes necessary to both plants and animals; yet too much copper is severely toxic to most green plants.

Life used and adapted to those elements and those concentrations available to it. When man started mining, using, and releasing these elements into the environment, the ecosystem was faced with hazards it had never before encountered, and to which it had, therefore, never adapted.

Toxicity of Cadmium

The toxicity of cadmium ion has been known for a long time. A local disease described by Japanese scientists, known as *itai-itai* (*ouch-ouch*), is now believed to have been caused by drinking river water contaminated by cadmium.

More recently, cadmium in drinking water has been suggested as a contributing factor towards heart disease. This is based on the observation that soft water may dissolve the cadmium, which is found with the zinc used to galvanize pipes. Thus, in hard water areas, there would be a protective layer of insoluble calcium salts lining the pipes. There does seem to be a higher incidence of heart disease in soft water areas.

Cadmium accumulates in the kidneys. Of the estimated 200–400 mg of cadmium normally ingested each day, about 3 g is retained in the body, with 1 g bound in the kidneys. By the time an individual is 30 years old, about 10 mg of Cd could have accumulated in the kidneys. This observation has led some investigators to suggest that Cd is implicated in hypertension.

An important source of Cd may be cigarette smoke. In comparing the total cadmium content of the kidneys, liver, and lungs of subjects at their death, non-smokers had a mean cadmium level of 6.63 mg as compared to 15.80 mg for smokers. Age and occupation were not significant variables, but number of cigarettes was significant. Much investigative work remains to be done to determine if the Cd in cigarette smoke contributes to health problems attributed to cigarette smoking.

The insoluble salt, cadmium sulphide is used in shampoo for the treatment of dandruff. Cadmium vapors are highly toxic, with maximum acceptable concentration in the air of 0.1mg/m. When Cd is heated to 400°C, the saturated vapor content is 8,000 mg/m. Thus wherever Cd is used at elevated temperatures, extreme precautions (respirators and ventilators) are mandatory.

Even brief exposure leads to a persistent cough and chest pains; these may be followed by pulmonary edema and death. An insidious feature is that symptoms may be delayed for some hours in dangerously high concentrations. Ingestion leads to almost immediate gastrointestinal poisoning, similar in its symptoms to food-poisoning. Kidney and liver damage may also be caused. After a long exposure to small amounts, cadmium may become concentrated in the gonads.

Because cadmium salts such as laurate, naphthenate, and stearate may be used in plastics manufacture, a report has set an upper limit of 2ppm for cadmium in plastics used for food containers.

Toxicity of Lead

Lead and its compounds have had a long medical history. Although recognized for centuries as being toxic, it has been widely used industrially, in food and beverage processing, and in medicine. *Plumbism* (lead poisoning) at one time was an occupational hazard in the manufacture of storage batteries and pottery. Many commercial painters became ill or died in the early twentieth century from the use of paints containing pigments made from lead. These same paints are now affecting children who ingest the paint chips.

Today, lead poisoning is a well-known hazard in our society. Indeed, it is becoming uncomfortably well recognized as sources of lead, ranging from older paint and automobile fumes to earthenware utensils, cocktail glasses, and moonshine whiskey, are continually being exposed.

Lead is a cumulative poison, the metal being stored in the solid portion of the body skeleton. For this reason, even after exposure has ceased, its excretion from the body may continue for years.

The symptoms of chronic poisoning are great weakness, constipation, severe abdominal cramps (lead colic), marked anemia and at times psychic manifestations and paralysis. In children a fatal encephalopathy due to cerebral edema is common.

The RBC's of a patient with lead poisoning show a peculiar stippling that is characteristic of this disease. The deposit of lead sulphide about the blood vessels near the margins of the gums sometimes causes a visible "lead line."

The prognosis is good in most cases of lead poisoning, although symptoms may persist for many years. Paralysis and mental symptoms in advanced cases may be permanent.

Treatment is directed towards relief of symptoms, removal of the cause and increased excretion of lead from the body. This is effected by means of chelating agents which combine with the metal to form a soluble and readily excreted substance. While lead may be considered a protein precipitant by combining with the cysteine sulfhydryl groups of protein, chronic lead poisoning manifests itself by inhibition of heme synthesis.

Toxicity of Mercury

Mercury and its compounds with few exceptions are highly poisonous to living organisms. Particularly in connection with finely-divided mercury metal, extreme care must be taken to avoid inhalation of and contact with the element. The chances of poisoning are increased because awareness of the presence of metal is reduced when it is finely divided. The fine grey mercury powder is easily generated when liquid mercury is rubbed against or agitated with grease, chalk, sugar, ether, and numerous other substances.

Mercury is absorbed through the gastro-intestinal tract and also through skin and lungs. After absorption, it circulates in the blood and is stored in the liver, kidneys, spleen, and bone. Elimination takes place only slowly in urine, feces, sweat, and saliva. Very mild exposure to the vapor may produce dryness of the throat and mouth, while the main symptoms of more prolonged exposure are inflammation of the mouth and gums, loosening of the teeth, tremors, and psychic disturbance.

This last effect (erethism) is characterized by loss of memory, lack of confidence, insomnia, and depression. In more severe cases, convulsive movements, paralysis, serious intestinal and urinary complications and death occur.

The maximum acceptable concentration of mercury vapor in air is 0.01 ppm. In laboratories, the danger arises chiefly because of the high vapor pressure of mercury at room temperatures which is about 20 times the m.a.c., and because long exposure to low concentrations produces a cumulative effect. There is also evidence that some individuals may become hypersensitive to mercury vapor.

Organo-mercury compounds are very hazardous not least because of their frequently high volatility. Their inhalation or absorption via the skin leads rapidly to nervous disorders, tremor, ataxia, difficulty in speech, constriction of the visual field and possibly psychosis. Alkyl-mercury derivatives present a special hazard because of their tendency to accumulate in the brain, where they cause irreversible damage.

Compounds of mercury, especially when soluble, are very toxic when ingested and less than 0.5g may prove fatal. Severe intestinal disturbances together with cardiac arrests are typical. Hg(II) fulminate, although seldom giving rise to poisoning, may induce dermatitis. Allergic behavior towards mercuric chloride is also known.

Mercury poisoning due to ingestion of contaminated grains and other foodstuffs can be prevented by proper food inspection, labeling, and if necessary, outright banning of mercurial fungicides. The toxic effects of mercury like those of lead and arsenic are due to its combining with protein sulfhydryl groups (*mercaptans*).

Under controlled conditions, mercury compounds have valuable pharmaceutical uses e. g; RHgX derivatives are widely used as diuretics and other organo--mercurials are effective in the treatment of syphilis. They are also highly toxic towards fungi and bacteria generally.

Toxicity of Arsenic

Arsenic compounds injure or destroy all cells and are known as protoplasmic poisons. Arsenic reacts with sulfhydryl groups of protein and simple molecules. Arsenicals have had a long history in medicine and in crime. Arsenic as a potassium arsenite solution has been used for leukaemia. There is some reason now to believe that inorganic arsenicals may be carcinogenic. More common than the macabre has been chronic arsenic poisoning from industrial and drinking waters. There has been the traditional belief that one can build up a tolerance to arsenic by ingesting small amounts daily. Whether or not this is true is debatable.

The toxicity of inorganic arsenic depends on its valence state and also on the physical and chemical properties of the compound in which it occurs. Trivalent (As^{3+}) compounds are generally more toxic than pentavalent (As^{5+}) compounds. It should be noted that animals are generally less sensitive than humans to the toxic effects of arsenic.

Symptoms of acute inorganic arsenic poisoning are vomiting, epigastric and abdominal pain, and diarrhoea. Muscle cramps and cardiac abnormalities have also been reported. Severe exposures can result in acute encephalopathy, stupor, convulsions, paralysis, coma, and even death.

Symptoms of chronic arsenic poisoning in humans are weakness and lassitude, loss of appetite and energy, loss of hair, hoarseness, and mental disorders.

CHAPTER 3

Biochemistry of Cell and Cellular Materials

1. Cations in Biological Systems

The cell is the fundamental unit of all living forms. Every manifestation of biological activity in all cells results from underlying chemical processes. Cells of various tissues and organs differ from one another chiefly in:

The nature and quantity of chemical substances present, The nature of the reactions in which the constituents participate, and The rates of these reactions.

The classification of cations in biological systems is given in the **Table III:**

Table III. Classification of Cations in Biological Systems

Na^+, K^+ Mg^{2+}, Ca^{2+} Fe, Cu	Alkali metals are: Charge carriers, mobile, O-Anion binding, weak complexes, very fast exchange.
	Mg and Ca: Structure formers and triggers, semi-mobile, O-anion binding, Moderately strong complexes, moderately fast exchange.
	Fe and Cu: Redox catalysts, static, N/S Ligands, strong complexes, no exchange.

The biological significance of Na^+, K^+, Mg^{2+}, Ca^{2+} cations is very different from that of the transition metals. Whereas the transition metals are strongly bound and immobile, the Group IA and IIA metals are weakly bound and mobile. These differences greatly affect the functions of the metals for whereas the cations of Cu, and Fe are redox catalysts, the Na, K, Mg, and Ca cations are concerned more in structural and transport (of ionic charge) roles than directly in catalysis. In addition and because of their mobility, their concentration in a given cell can be controlled by metabolism so that cell activity is proportional to the free cation concentration.

A major chemical procedure which is used to follow the chemistry of Na, K, Mg, Ca, Fe, and Cu etc; in biological systems is that of strict isomorphous replacement. It is known from mineralogy that cations substitute for one another on the basis of similarities in radii and in stereo-chemical demands. Ionic charge is important but replacement of M^{2+} by M^+ and M^{3+} by M^{2+} is often possible.

2. Accumulation of Na, K, Mg, Ca

It has been known for a long time that the concentration of free potassium inside is much higher than that outside cells whereas the level of sodium is generally much lower. Such a separation requires constant expenditure of energy and a selective pump which recognizes the difference between the two cations. The minimum energy used per unit time is proportional to the ion gradient, and the flux required to maintaining the gradient. Ignoring the flux problem, the free energy requirement is proportional to ΔG_1:

$$\Delta G_1 = RT \ln \frac{[K^+] \text{ in}}{[K^+] \text{ out}} + RT \ln \frac{[Na^+] \text{ Out}}{[Na^+] \text{ in}} \qquad (3.1)$$

For a series of cells which have the same external cellular environment then the relative free energy required is:

$$\Delta G_2 = RT \ln \frac{[K^+] \text{ in}}{[Na^+] \text{ in}} \qquad (3.2)$$

It has also been known that intra-cellular Mg concentrations are often higher than extra-cellular concentrations but that in-cell calcium is maintained at a rather low level.

In 1967 it was suggested by Williams and Wacker that cells operated a general pumping of these bivalent cations much as they pumped univalent cations.

For a series of cells which have the same cellular environment the energy required to generate the concentration gradient is proportional to $\Delta G3$:

$$\Delta G_3 = RT \ln \frac{[Mg^{2+}] \text{ in}}{[Ca^{2+}] \text{ in}} \qquad (3.3)$$

The pumping of the cations is brought about by the hydrolysis of adenosine triphosphate (ATP) through enzymes called Na^+/K^+ ATP- ases for Na^+ and K^+ pumping, and Ca^{2+} ATP-ases for Ca^{2+} pumping, which are in the outer cell membrane. Thus both processes are regulated by the energy supply, the concentration of ATP.

Interestingly, the concentration of the cations Mg^{2+} and K^+ is paralleled by an anion Concentration gradient:

$$\Delta G_4 = RT \ln \frac{[\text{Phosphate}] \text{ in}}{[\text{Chloride}] \text{ in}} \qquad (3.4)$$

A very crude measure of the activity of a cell is provided by bound internal phosphate for it requires energy to condense phosphate with other inorganic and organic residues. It follows that one should inspect [P], [Cl⁻], [Na^+], [K^+], [Mg^{2+}] and [Ca^{2+}] changes together.

The erythrocytes of dog, the cat, and certain sheep give a low ratio [K^+] in/ [Na^+] in but quite a high ratio [Mg^{2+}] in / [Ca^{2+}] in. Among particular sheep the univalent ionic concentration gradient [K^+] in / [Na^+] in of the red blood cells (RBC's) is generally transmitted from generation to generation.

Even the different muscles of the body can be classified on the basis of their inorganic contents, which in turn maybe related to total ATP-ase activities. Moreover, the onset of many diseased conditions e. g; dystrophy and eye cataracts, is accompanied by changes in all three concentration ratios towards high [Na^+], [Ca^{2+}], and [Cl⁻]. All this points to the general conclusion that there is a strictly controlled separation of K^+ from Na^+, and of Mg^{2+} from Ca^{2+} in all living cells.

The ability to reject Na^+ and Ca^{2+} and to accumulate or maintain K^+ and Mg^{2+} have led to a striking functional differentiation of these cations in biological systems. In bacteria, the separation of the elements could then be evolved into control systems associated with replication. In more complex living systems, which have protected environments, the cell membranes become more permeable. Thus the Na^+ and K^+ gradients set up in the higher organisms established membrane potentials which have permitted the development of nerve, muscle, and brain.

CHAPTER 4

Biochemistry of Essential Metal Ions

Sodium

Sodium is the principal cation in the extra-cellular fluid (both interstitial and vascular fluids) compartments. It is absolutely essentially to the normal health and growth of living things. This ion is responsible for maintaining normal hydration and osmotic pressure.

The importance of sodium in human nutrition, as exemplified by the need of the salt, has been recognized since the very early days of man. Man and animals, who live primarily on fruits, grains and vegetables receive an adequate level of potassium from these sources, but require additional salt and will experience a sodium deficiency if they do not get it. On the other hand, men and animals, who live on meat, milk, and other foods derived from animals, receive sodium into their bodies from these sources, and do not need the additional salt.

Sodium deficiencies result when sodium ion is lost from the body by excessive sweating and other loss of body fluids and is not replenished. This gives rise to such symptoms as thirst, nausea, muscle cramps, mental disturbances, and even result in death.

Normally more than adequate amounts of sodium are contained in the daily diet with nearly complete absorption from the intestinal tract. Excess sodium is excreted by the kidneys, which make them the ultimate regulator of the sodium content of the body. Approximately 80–85% of the sodium in the glomerular filtrate is reabsorbed. This reabsorption is under hormonal control that is still not completely understood.

Conditions causing *hyponatremia* (low serum sodium level) are: Extreme urine loss, such as seen in diabetes inspidus (a disease of pituitary origin as contrasted with diabetes mellitus, which is caused by deficient insulin secretion by the B-cells of the islets of Langerhans in the pancreas); metabolic acidosis, in which sodium is excreted; Addison's disease, with decreased excretion of the anti-diuretic hormone, aldosterone; Diarrhoea and vomiting; Kidney damage.

Hypernatrmia (increased serum sodium level) is found in: Hyperadrenalism (Cushing's Syndrome) with increased aldosterone production; Severe dehydration; Certain types of brain injury; and Excess treatment with sodium salts.

There is a good correlation between sodium content (as sodium chloride) of the tissues and hypertension. If for some reason, the body is unable to eliminate sodium and the concentration starts to increase, water is retained in the tissues to maintain osmotic balance. Oedema results and outwardly, the patient can take on a puffy appearance with swelling, particularly of the lower extremities. The build-up of fluids puts on added burden on the heart which may be aggravated if the heart is also diseased. Treatment includes low salt diets, cardio-tonic drugs, and combinations of each. In temporary conditions such as pregnancy, elimination of salt and highly salted foods will greatly reduce the Oedema and concurrent problems. Sodium –free salt substitutes (Neocurtasal, Co-salt) can be used to enhance the flavour of food. Co-salt is a blend of Choline, KCl, NH_4Cl and tricalcium phosphate.

The association of sodium in body metabolism is given in **Table IV:**

Table IV. Na⁺ and Body Metabolism

Total Amount in Human Body = 1.8g/Kg, RDA* = 3-5 g	
Best Food Sources:	Table salt, salty foods, Animal Foods, Milk, Baking Soda, Baking powder. Some vegetables.
Absorption and metabolism:	Readily absorbed, Extra-cellular, Excreted in urine, And sweat, Aldosterone Increases reabsorption in renal tubules.
Principal metabolic functions:	Buffer Constituent Acid-base balance Water balance, Osmotic pressure CO_2 transport, Cell membrane Permeability, Muscle irritability.
Clinical manifestations of deficiency:	Dehydration, Acidosis, Tissue atrophy, Edema, Hypertension
RDA*= Recommended Dietary Allowance	

Potassium

Potassium is the major intra-cellular cation present in a concentration ~23 times higher than the concentration of potassium in the extra-cellular fluid compartments. This concentration differential is maintained by an active transport mechanism. During transmission of a nerve impulse, potassium leaves the cell and sodium enters the cell. It is currently thought that an active transport mechanism re-establishes the concentration differential after transmission of the nerve impulse. This active transport mechanism has been called the Sodium Potassium Pump. It is the difference in the Na: K ratio in intra-cellular fluids compared with the ratio in the extra-cellular fluids that determines the various electro-physiological functions in higher animals.

Potassium in the diet is rapidly absorbed. Any excess potassium is rapidly excreted by the kidneys. Fortunately man maintains an adequate level of potassium by means of a normal diet, particularly one that includes an adequate proportion of vegetables and plant-derived foods. Potassium salts have been used for their diuretic action because of the efficient excretion of potassium by the kidneys, since a certain volume of fluid (urine) will be excreted in order to keep the potassium salt in solution.

Both elevated and low potassium levels can be serious to the patient. *Hypokalemia* causes changes in myocardial function, flaccid and feeble muscles, and low blood pressure. It can occur from vomiting, diarrhoea, burns, hemorrhages, diabetic coma, intra-venous infusion of solutions lacking in potassium (a dilution effect), overuse of thiazide diuretics, and alkalosis. The latter occurs due to the movement of potassium into the cell as protons move out of the cell into the proton-deficient extra-cellular fluid. Notice that, in this situation, it is possible for a low serum potassium level and elevated potassium level to occur concurrently.

Hyperkalemia is less common and usually occurs during certain types of kidney damage. If the kidney is functioning properly, the body can eliminate excess potassium readily. In certain acidotic conditions, interference with the sodium and potassium proton exchange can result in potassium retention.

The heart appears to be particularly sensitive to potassium concentrations. In hypokalemia there are alterations in the electrocardiogram (ECG) and distinct histological changes in the myocardium. An increase in potassium levels also results in changes in the ECG and causes the heart muscle to become flaccid with possible cessation of heart beat (potassium arrest). It is thought potassium may be displacing calcium in the cardiac muscle, since a decrease in calcium will produce a similar pattern in the heart muscle and may explain why calcium gluconate is effective in hyperkalemic conditions.

Because potassium is the major intra-cellular cation, serum potassium levels may not be a sensitive enough measure of the body's potassium levels. There is a current effort at developing whole body counts of potassium by measuring levels of K^{40}.

Potassium had been recognized as an essential nutrient for the growth of plants by Von Liebig as early as 1840. Potassium is taken into plants through the root system by absorption from the soil. The exact mechanism is not thoroughly established, but it is known that potassium is directly involved in photosynthesis and respiration. It is also known that potassium is absorbed selectively over other cations found in the soil, such as calcium or magnesium. Potassium deficiencies in plants show up as leaf curling, yellow or brown spots, and drastically retarded root growth.

The association of potassium in body metabolism is given in **Table V:**

Table V. K⁺ and Body Metabolism

Total Amount in human body = 2.6g/Kg, RDA (Per Day) = ~1.5–4.5g	
Best Food Sources:	Vegetables, Fruits, Whole grains Meat, Milk, Legumes.
Absorption & Metabolism:	Readily absorbed, Intra-Cellular, Excreted by Kidney.
Principal Metabolic functions:	Buffer constituent, Acid-base balance, Water balance, CO_2 transport, Membrane transport, Neuro-muscular Irritability.
Clinical manifestations of deficiency:	Acidosis, Renal damage

Calcium

The biochemistry of calcium is very complex. 99% of body calcium is found in the bones. The remaining calcium is largely found in extra-cellular fluid compartments. The biochemical functions of calcium, as well as the hormonal control of serum calcium levels, are very complicated, with many of the details yet to be elucidated.

Calcium is absorbed from the upper part of the small intestine where the intestinal contents are still acidic, and it exists as ionized water-soluble salts.

As the intestinal contents become neutral to alkaline, calcium is precipitated as the dibasic phosphate ($CaHPO_4$), carbonate, oxalate, and sulphate salts, and as insoluble calcium soaps.

The actual absorption of calcium across the intestinal membranes is controlled by the parathyroid hormone and a metabolite of vitamin D. It is currently believed that cholecalciferol (vitamin D_3) is hydroxylated at the C-25 position in the liver and then at the C-1 position in the kidneys. This activated metabolite, 1, 25-dihydroxycholecalciferol, may function as a gene-activator, causing the synthesis of a calcium-binding protein which transfers the calcium cation across the intestinal wall.

The intestinal absorption and serum level of calcium are highly influenced by phosphate. Calcium absorption and distribution are also under a complex hormonal control; parathyroid hormone (PTH) and the recently discovered calcitonin (thyrocalcitonin). Removal of the parathyroid glands causes a serious tetany to develop due to a sharp drop in serious calcium levels and a rise in phosphorus levels. Administration of PTH raises the blood calcium, lowers blood phosphorus, and increases the elimination of both in the urine.

Functionally 99% of all body calcium is supportive, being found in bone as hydroxyapatite. The remaining ionic calcium is involved in neuro-hormonal functions, blood clotting, muscle contraction, and possibly in other biochemical processes. The deleterious effects of hyper-potassemia on the heart may be due to excessive potassium displacing calcium from the cardiac muscle.

Hypercalcemia is found in hyperparathyroidism, hyper-vitaminosis D, and some bone neoplastic diseases. Symptoms include fatigue, muscle weakness, constipation, anorexia (loss of appetite), and cardiac irregularities. If the condition persists, calcium salts may be deposited in the kidneys and blood vessels. There are many methods of reducing the intestinal absorption of calcium, ranging from the precipitation of calcium as insoluble sulphate or phosphate salts to complexation with EDTA. Recent reports indicate that cellulose phosphate is effective in reducing intestinal calcium hyperabsorption.

Hypocalcemia can be caused by hypoparathyroidism, vitamin D deficiency, osteoblastic metastatis (spreading bone cancer), *Cushing's Syndrome* (hyperactive adrenal cortex), acute pancreatitis, and acute hyperphosphatemia. If the serum calcium levels fall enough, hypocalcemic tetany can result.

Associated with the above condition are disorders in bone metabolism. Bone is a dynamic tissue involving constant exchange of calcium and phosphate ions with the body fluids. Much of this exchange is under hormonal control.

Bone in addition to providing structural support, is also a storage tissue for calcium. At night, when a person goes for 12 hours or more without food intake, resorption of the bone occurs in order to maintain blood calcium levels.

Calcium is an essential component of plants. It stimulates the growth of root hairs. Cell walls formed in its absence are weak. Lack of it interferes with the activity of the chloroplasts and retards the movement and utilization of carbohydrates and amino acids.

The association of calcium with body metabolism is given in **Table VI:**

Table VI. Calcium and Body Metabolism

Total amount in human body = 22g/Kg RDA = 100mg	
Best Food Sources:	Milk, Milk Products, Fish
Absorption &Metabolism:	Poorly absorbed according to body need; Absorption aided by Vitamin D, lactose.
Principal Metabolic Functions:	Formation of apatite in bones, teeth; blood clotting, cell-membrane permeability, neuro-muscular irritability.
Clinical manifestations of deficiency:	Rickets (child), Poor growth, osteoporosis (adult).

Magnesium

Ionic magnesium and calcium are present in most soils; they constitute up to 5 and 25% respectively of the dissolved solids of most surface waters. Therefore, they are available to all forms of plant and animal life, and fortunately so because they are essential elements. Chlorophylls a and b are Mg-chelates-$C_{55}H_{(70, 72)}N_4O_{(6, 5)}$ Mg important in photosynthesis.

Both calcium and magnesium serve as regulators of phosphate transport and deposition in plants and animals. Bones and teeth are particularly high in these elements. There is some evidence that ratios of calcium and magnesium to sodium and potassium in blood are significant with reference to problems of the circulatory system.

Magnesium is the second most plentiful cation in the intra-cellular fluid compartment and the fourth most abundant cation in the human body.

50% of total body magnesium is combined with calcium and phosphorus in bone. It is an essential component of many of the enzymes involving phosphate metabolism which also require adenosine triphosphate. Magnesium is also apparently indispensable for protein synthesis and for the smooth functioning of the neuro-muscular system.

Negative magnesium balance is more widespread than most people generally realize. Causes include malnutrition, dietary restrictions, gastro-intestinal diseases, chronic alcoholism, faulty absorption or utilization, and parathyroid hormone (PTH) imbalances. The clinical significance of magnesium depletion varies widely both in symptomatology and in biochemistry, depending upon the overall health of the individual.

Symptoms of magnesium deficiency include personality changes after depletion of three or four month's duration, failure to gain weight properly, and cardiac disturbances. Hypomagnesemia and alkalosis have been correlated with the withdrawal symptoms of the chronic alcoholic.

Magnesium cation has a definite pharmacological action. Magnesium salts when injected intravenously or intramuscularly have a powerful general anaesthetic action which resembles that produced by chloroform. Soluble magnesium salts (usually magnesium sulphate) have been used as central nervous system depressants in obstetrics, convulsion states, and for the symptoms of tetanus.

Magnesium ion is not readily absorbed from the gastro-intestinal tract because its absorption is retarded by alkaline media. Most of the absorption takes place in the acid medium of the duodenum. Due to the slow absorption of magnesium ions, a saline laxative action occurs upon the ingestion of any water-soluble magnesium compound.

The association of magnesium with body metabolism is given in **Table VII:**

Table VII. Magnesium and Body Metabolism

Total Amount in Human Body = 0.5g/kg	RDA = 400–420mg for men, 310–320 for women
Best Food Sources:	Chlorophyll, nuts, legumes. Whole grains.
Absorption & Metabolism:	Absorbed, competes with calcium for transport.

Principal Metabolic Functions:	Co-factor for PO_4^- transport enzymes, Constituent of bones and teeth. Decreases Neuro-muscular irritability
Clinical manifestations of deficiency:	Mg-Conditioned deficiency, Muscular tremor, choreiform movemets, Confusion, Vasodilation, Hyperirritability.

Biochemistry of Iron

Iron is an extremely important constituent of blood and tissues of the animal body. Iron is present in some form wherever respiration occurs in higher animals. It is essential to the elementary metabolic processes in the cell. In the respiratory chain, iron functions as an electron carrier. Iron is responsible for the transport of molecular oxygen in higher organisms. Both of these functions depend on the ability of iron to exist in coordination compounds in different states of oxidation and bonding.

Most of the iron found in the body is associated with three types of proteins: Heme proteins, Non-heme proteins, and Fe storage and / or transport proteins. The chief heme proteins are: Hemoglobin, (b) Myoglobin, (c) Cytochromes, and (d) Oxidative enzymes.

Haemoglobin

Haemoglobin (frequently abbreviated as **Hb**) is the iron-containing oxygen-transport metalloprotein in the red cells of the blood in mammals and other animals. Haemoglobin in vertebrates transports oxygen from the lungs to the rest of the body, such as to muscles, where it releases the oxygen-load. Haemoglobin also has a variety of other gas-transport and effect-modulation duties, which vary from species to species, and which in invertebrates may be quite reverse.

The name *haemoglobin* is the concatenation of *heme* and *globin*, reflecting the fact that each subunit of haemoglobin is a globular protein with an embedded heme group; each heme group contains an iron atom, and this is responsible for the binding of oxygen. The most common types of haemoglobin contain four such subunits, each with one heme group.

Mutations in the genes for the haemoglobin protein in humans result in a group of hereditary diseases termed the *hemoglobinopathies*, the most common members of which are sickle-cell disease and thalassemia. Historically, in human medicine, haemoglobinopathies were the first diseases to be understood in mechanism of dysfunction.

Structure of Haemoglobin

The haemoglobin molecule is an assembly of four protein subunits. Each subunit is composed of a protein chain tightly associated with a non-protein heme group. Each individual protein chain arranges in a set of alpha-helix structural segments connected together in a 'myoglobin fold" arrangement, so called because this arrangement is the same folding motif used in the heme/globin proteins. This folding pattern contains a pocket which is suitable to strongly bind the heme group.

A heme group consists of an iron atom held in a heterocylic ring known as a *Porphyrin*. This iron atom is the site of oxygen binding. The iron atom is bonded equally to all four nitrogens in the center of the ring, which lie in one plane.

Two additional bonds perpendicular to the plane on each side can be formed with the iron to form the fifth and sixth positions, one connected strongly to the protein, the other available for binding of oxygen. The iron atom can either be in the Fe^{2+} or Fe^{3+} state, but ferri-haemoglobin (Methaemoglobin) (Fe^{3+}) cannot bind oxygen. Hb exists in two forms, a taut (tense) form (T) and a relaxed form (R). Various factors such as low pH, high CO_2 at the level of the tissues favour the T form which has low oxygen affinity and releases oxygen in the tissues.

Structure of Heme Group is given In Fig. I.

The Fe^{2+} in the hemoglobin may exist in either a high-spin (deoxygenated) or low-spin (oxygenated) state, according to population of the iron (II) d-orbital structure with its 6 available d electrons, as understood in crystal field theory. With the binding of an oxygen molecule as a sixth ligand to iron, the iron (II) atom finds itself in a *octahedral field* (defined by the six ligand points of the four porphyrin ring nitrogens, the histamine nitrogen, and the O_2). In these circumstances, with strong field ligands, the five d-orbitals (these are the "3d" orbitals of the iron) undergo a splitting in energy between two of the d-orbitals which point directly in the direction of the ligands (d_z^2 and $d_{x}^2 - d_{y}^2$ orbitals, hybridized in these circumstances into two e_g orbitals), and three of the d-orbitals which are pointed in off-directions (the d_{xy}, d_{xz}, d_{yz} hybridized in these circumstances into three t_{2g} orbitals).

CHEMISTRY AND OUR DAILY EXISTENCE

A heme group consists of an iron atom held in a heterocylic ring known as a *Porphyrin*. This iron atom is the site of oxygen binding. The iron atom is bonded equally to all four nitrogens in the center of the ring, which lie in one plane. Two additional bonds perpendicular to the plane on each side can be formed with the iron to form the fifth and sixth positions, one connected strongly to the protein, the other available for binding of oxygen. The iron atom can either be in the Fe^{2+} or Fe^{3+} state, but ferrihemoglobin (Methaemoglobin) (Fe^{3+}) can not bind oxygen.

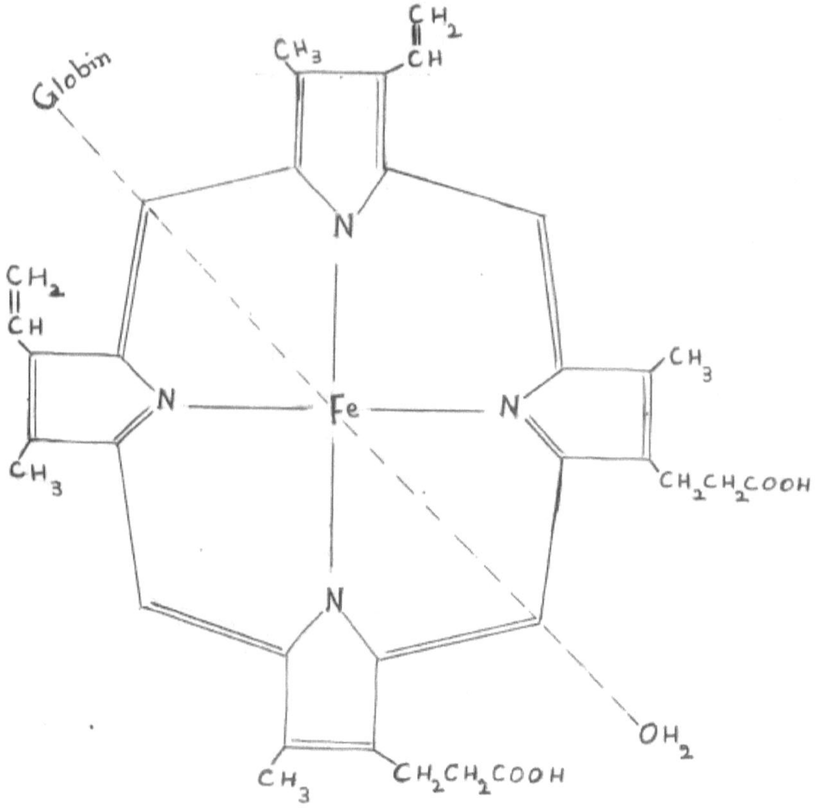

Fig I Heme group

When oxygen is bound to Fe^{2+} in heme, all six d-electrons are forced into the three lower energy t_{2g} orbitals, where they must all be paired. This produces the "low-spin" state of oxy- haemoglobin. The sharp high –energy of transition between the t_{2g} and empty *eg* states of d-orbital electrons in oxy-haemoglobin is responsible for the bright red colour of the substance. When oxygen leaves, the Fe is allowed to move out of the porphyrin ring plane, away from its five ligands toward the empty space formerly occupied by the O, and in these circumstances *eg* orbital energies drop and *t2g* electrons move into them. This causes the iron atom to expand and increase its net spin, as d-orbitals become populated with unpaired electrons. In these circumstances, the absorption spectrum becomes broader, with smaller transition levels, producing dark colour of deoxyhaemoglobin.

In adult humans, the most common haemoglobin type is a tetramer (which contains 4 subunit proteins) called haemoglobin A, consisting of two a and two B subunits non-covalently bound, each made of 141 and 146 amino acid residues respectively. This is denoted as a B. The subunits are structurally similar and about the same size. Each subunit has a molecular weight of about 16,000 daltons, for a total molecular weight of the tetramer of about 64,000 daltons. Haemoglobin A is the most intensively studied of the haemoglobin molecules.

The four polypeptide chains are bound to each other by salt bridges, hydrogen bonds, and hydrophobic interaction. There are two kinds of contacts between the ά and Б chains: $ά_1Б_1$ and $ά_1B_2$.

Oxygen-Binding Capacity of Haemoglobin

Haemoglobin's oxygen-binding capacity is decreased in the presence of carbon monoxide because both gases compete for the same binding sites on haemoglobin, carbon monoxide binding preferentially in place of oxygen.

Carbon dioxide occupies a different binding site on the haemoglobin. Through the enzyme carbonic anhydrase, carbon dioxide reacts with water to give carbonic acid, which decomposes into bicarbonate and protons:

$$CO_2 + H_2O \rightarrow H_2CO_3 \rightarrow HCO_3 + H^+ \qquad (4.1)$$

Hence blood with high carbon dioxide levels is also lower in pH (more acidic). Haemoglobin can bind protons and carbon dioxide which causes a conformational change in the protein and facilitates the release of oxygen. Protons bind at various places along the protein, and carbon dioxide binds at the alpha-amino group forming carbamate. Conversely, when the carbon dioxide levels in the blood decrease (i.e; in the lung capillaries), carbon dioxide and protons are released from haemoglobin, increasing the oxygen affinity of the protein. This control of haemoglobin's affinity for oxygen by the binding and release of carbon dioxide and acid, is known as the Bohr effect.

The binding of oxygen is affected by molecules such as carbon monoxide (CO) (for example from tobacco smoking, cars and furnaces). CO competes with oxygen at the same binding site. Haemoglobin binding affinity for CO is 200 times greater than its affinity for oxygen, meaning that small amounts of CO dramatically reduces haemoglobin's ability to transport oxygen.

When haemoglobin combines with CO, it forms a very bright red compound called carboxy-haemoglobin. When inspired air contains CO levels as low as 0.02%, headache and nausea occur; if the CO concentration is increased to 0.1%, unconsciousness will follow. In heavy smokers, up to 20% of the oxygen-active sites can be blocked by CO.

In similar fashion, haemoglobin also has competitive binding affinity for cyanide (CN^-), Sulfur monoxide (SO), nitrogen dioxide (NO_2), and sulfide (S^{2-}), including hydrogen sulfide (H_2S). All of these bind to iron in heme without changing its oxidation state, but they nevertheless inhibit oxygen-binding, causing grave toxicity.

Role in Disease

Decreased levels of haemoglobin, with or without an absolute decrease of RBC;s, leads to symptoms of anaemia. Anaemia has many different causes, although iron deficiency and its resultant iron deficiency anaemia are the most common causes in the Western world. As absence of iron decreases heme synthesis, red blood cells in iron deficiency anaemia are *hypochromic* (lacking the red haemoglobin pigment) and *microcytic* (smaller than normal). Other anaemias are rare. In hemolysis (accelerated breakdown of RBC's), associated jaundice is caused by the haemoglobin metabolite bilirubin, and the circulating haemoglobin can cause renal failure. Mutations in the globin chain are associated with the haemoglobinopathies, such as sickle-cell disease and thalassemia. There is a group of genetic disorders, known as the *porphyrias* that are characterized by errors in metabolic pathways of heme synthesis. King George III of the United Kingdom was probably the most famous porphyria sufferer.

The association of iron with body metabolism is given in **Table VIII:**

Table VIII. Iron and Body Metabolism

TOTAL amount in human body = 75mg/kg	RDA = 8mg for males, 18mg for females
Best Food Sources:	Liver, Meat, egg-yolk, Green leafy vegetables, Ascorbic acid, whole grains, Enriched bread, and cereals.

Absorption & metabolism:	Absorbed according to body need. Aided by HCl, Ascorbic acid.
Principal metabolic functions:	Constituent of haemoglobin, myoglobin, catalase, ferredoxin, cytochromes, Electron transport, enzyme co-factor.

Myoglobin

Myoglobin is closely related to haemoglobin. The familiar colour of muscle is due to the presence of myoglobin within muscle cells. The oxygen of myoglobin (delivered by haemoglobin coming from the lungs to muscles) may be considered as a store rather than as part of the transport mechanism In muscle at rest, oxygen probably remains fixed to myoglobin. During the contractile phase, when the demand for oxygen is maximal, oxygen dissociates from myoglobin and is used for intra-cellular oxidations. In diving mammals, the myoglobin content of muscles is particularly high and probably enables the animals to stay under water for long periods. Myoglobin however, does not exhibit the Bohr Effect (pH dependence), presumably indicating that the heme moiety is not linked to an imidazole residue as it is in the haemoglobin.

Myoglobin must have a greater affinity for oxygen than haemoglobin in order to affect the transfer of oxygen at the cell. The equilibrium constant for myoglobin-oxygen complexation is given by the simple equilibrium expression:

$$K_M = \frac{[MbO_2]}{[Mb][O_2]} \qquad (4.2)$$

Myoglobin is largely converted to oxymyoglobin even at low oxygen concentrations as occur in the cells.

The equilibrium constant for the formation of oxy-haemoglobin is somewhat more complicated. The expression in the range of physiological importance in the tissues is:

$$K_H = \frac{[MbO_2]}{[Hb][O_2]^{2.8}} \qquad (4.3)$$

The 2.8 exponent for oxygen results from the fact that a single haemoglobin molecule can accept four oxygen molecules and *the binding of the four is not independent.*

The net result is that at low oxygen concentrations haemoglobin is less oxygenated and at high oxygen concentrations haemoglobin is more oxygenated than if the exponent were 1. This effect favours oxygen transport since it helps the haemoglobin get saturated in the lungs and deoxygenated in the capillaries.

The Oxygen binding curves for Hb and Mb are given in Fig. II

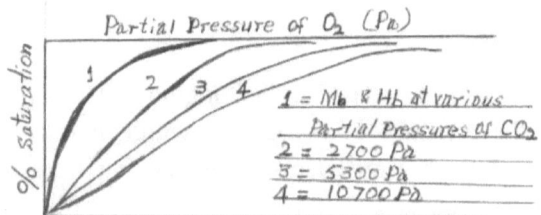

Fig.II. Oxygen binding curves for Hb and Mb.

The iron in myoglobin and hemoglobin is in the +2 oxidation state. The oxidized form containing iron (III), called metmyoglobin and methemoglobin, *will not bind oxygen.* It is of special interest to note that free heme is immediately oxidized in the presence of oxygen and water and thus rendered useless for oxygen transport:

$$\text{Heme [Fe}^{II}\text{]} \xrightarrow[\text{water}]{O_2} \text{Hematin [Fe}^{III}\text{]} \quad (4.4)$$

In a biological system this would be fatal. The stability of heme (FeII) in myoglobin and hemoglobin is a result of the globin or protein portion of the molecule. Myoglobin has a molecular weight of 17, 000, of which most is a protein chain folded about the heme reducing access to iron and simultaneously producing a hydrophobic environment. This steric and chemical control allows access to and coordination by an oxygen molecule but does not allow the simultaneous presence of oxygen and one or more water molecules, which seems to be necessary for electron transfer.

Hemoglobin may be considered an approximate tetramer of myoglobin. It has a molecular weight of 64,500 and contains four heme groups bound to four protein chains. The differences between Hb and Mb in their behaviour towards oxygen are related to the structure and movements of the four chains. If the tetrameric Hb is broken into dimmers or monomers, these effects are lost. Upon oxygenation of hb, two of the heme groups move about 100 pm toward each other while two others separate by 700 pm. These movements seem responsible for the cooperative effects observed. In addition, Hb has a channel, 2000 pm wide, lined with polar groups. Protonation or deprotonation of these polar groups with changes in pH can contribute to the Bohr effect.

Perutz has suggested a mechanism to account for the cooperativity of the four heme groups in Hb. Basically it is founded on the idea that an interaction between an oxygen molecule and a heme group can effect the position of the protein chain attached to it, which in turn affects the other protein chains through hydrogen bonds. It has been dubbed the Rube Goldberg effect after the marvelous mechanisms consisting of ropes, pulleys, and levers in Goldberg's cartoons.

The key or "trigger" in the Perutz mechanism is the high-spin Fe (II) atom in an unbound (i. e; Oxygenless) heme. The radius of Fe is 78 pm. Perutz estimates that the Fe N bond distance in heme must therefore be 218 pm. Since there is only room for a bond length of about 200-205 pm, the iron atom must sit about 80 pm above the plane of the heme group. We know that coordination of the iron by oxygen results in spin-pairing of electrons. Low-spin Fe (II) is 17 pm smaller than high-spin Fe (II). The high-spin-low-spin "trigger" was first suggested by Hoard and is an interesting example of how a simple "inorganic" change, i. e; high-spin state to low-spin state, can be responsible for a highly important biological function.

CHAPTER 5
Cytochromes

1. Cytochromes

Cytochromes are heme proteins, found in both plants and animals, which serve as electron carriers. There are a number of these compounds, at least four of which occur in the common mitochondrial electron transport system. Owing to the differences of structure of different cytochromes, they differ in reactivity, particularly in their ability to accept and to donate electrons.

The reactions of the electron system are catalytic in nature. The compounds are used over and over again, alternately undergoing oxidation and reduction, so that the net effect is the oxidation of the organic substrate, the reduction of oxygen, and the synthesis of ATP. The important oxygen-activating enzyme is the cytochrome oxidase.

Electron transfer among the cytochromes, may be represented as:

$$\underset{\text{cytochrome}}{Fe^{++}} + \underset{\substack{\text{cytochrome}\\\text{oxidase}}}{Fe^{+++}\ OH^-} \rightarrow \underset{\text{cytochrome}}{Fe^{+++}\ OH^-} + \underset{\substack{\text{cytochrome}\\\text{oxidase}}}{Fe^{++}} \quad (5.1)$$

the terminal event, oxidation of cytochrome oxidase by O, may be represented as:

$$2Fe^{++} + \tfrac{1}{2}O_2 + H_2O \rightarrow 2Fe^{+++}\ OH^- \quad (5.2)$$

The active centre of the cytochromes is a heme group. It consists of a porphyrin ring chelated to an iron atom. The oxidation state of the iron may be either +2 or +3, and the importance of cytochromes lies in their ability to act as redox intermediates in electron transfer. They are present not only in the chloroplasts for photosynthesis but also in mitochondria to take part in the reverse process of respiration. The importance of cytochrome c in photosynthesis and respiration indicates that it is probably one of the oldest (in terms of evolutionary history) of the chemicals involved in biological processes.

Poisoning by cyanide or by sulfide anions, results from formation of complexes between the ferric iron of cytochrome oxidase and these anions. Methemoglobin (Met Hb contains ferric iron, formed by oxidation of ferrous iron of Hb) but not Hb forms similar complexes.

2. Cyanide Poisoning

The cyanide poisoning—-the instant killing used by suicide bombers, is briefly described below:

The structure and mechanism of haemoglobin has been explained on the basis of Perutz mechanism. The cooperativity of the heme groups is very real, and the "trigger" for the T $\leftrightarrow$ R inter-conversion *must* be something connected with the coordination of the oxygen to the iron. Therefore the Perutz mechanism involving Hoard's high-spin $\leftrightarrow$ low-spin equilibrium is the only one proposed that can account for it in a realistic way. It is interesting to note how the heme group acts as an "amplifier."

The change in configuration of the protein chains from the T (tense or deoxy) to the R (relaxed or oxy) form, triggered by the change of spin state of the iron atom, results in an estimated difference of 12–14 k J / mole in oxygen affinity by the two forms. This fundamental difference in the energetics of oxygen binding is responsible for the cooperativity of the four heme groups in haemoglobin. The reduced affinity of the T form is nature's device that makes it possible for haemoglobin to push the oxygen molecule off (R $\leftrightarrow$ T) in the tissues and transfer it to myoglobin.

Heme is an iron (II) complex of porphyrin, in which the iron atom has a coordination number of six. It is bonded to four Nitrogens of pyrrole rings in a plane and to globin perpendicular to this plane via a nitrogen atom of the imidazole group in histidine. The sixth position is probably occupied by a water molecule. This configuration permits the reversible combination with molecular oxygen at this sixth coordination site and it is by this mechanism that the RBC's of the blood carry oxygen from one part of the body to another. Haemoglobin and myoglobin contain high-spin iron (II) in the absence of a sixth ligand. Coordination by oxygen results in diamagnetic oxy-haemoglobin and oxy-myoglobin. This change in spin state is somewhat puzzling since molecular oxygen is not expected to be a particularly strong-field ligand. It is probable that the heme molecule is probably "tuned" by its substituents to make this spin pairing more likely. The exact molecular arrangement of oxygen in oxy-myoglobin and oxy-haemoglobin has been of great interest in understanding the chemistry of oxygen transport and storage. Fortunately it has been found from X-ray crystallographic analysis that the O_2 molecule coordinates in the bent geometry.

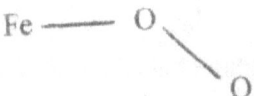

The vacant sixth coordination site can bind other ligands instead of oxygen. Strongly-bonding ligands such as cyanide ion CN^- are favoured. Similarly carbon monoxide, sulfur and phosphorus containing ligands, and other soft ligands bind more tightly than the weakly-binding oxygen and in the presence of large amounts of these, the Hb is tied up and unavailable for oxygen transport.

The formation of cyanide complexes is restricted almost entirely to the transition metals of the d block and their near neighbours Zn, Cd, and Hg.

The greater stability of such complexes is due to the following characteristics of the cyanide ion:

1. It occupies the top position in the spectrochemical (or Fajans-Tsuchida) series of ligands:

   $I < H_2O < C\underline{N} < \underline{C}N$ (the binding atoms are underlined).

   Cyanide ion is a ligand of the largest crystal field and in terms of Crystal Field Theory (CFT), it must be held most firmly.

   In fact, there is a good parallelism between strength of bonding and spectrochemical series. Jorgensen in 1962 quantified the spectrochemical series relative to water as a standard ligand with a field factor (f) of 1.00. values range from about 0.7 for weak-field bromide ions to about 1.7 for the very strong-field cyanide ion.

2. Cyanide ion gives rise to large *nephelauxetic effects*. The effect of ligands in expanding the d electron clouds of the metal has been named nephelauxetic ("cloud expanding") effect. The increase in the size of orbitals results from the combination of orbitals on the metal and ligand to form larger molecular orbitals through which the electrons can move. Ligands such as CN^- which can delocalize the metal electrons over a large space are most effective in this manner.

3. Cyanide ion produces a strong *trans effect*. The trans effect may be defined as the labilization of ligands trans to other trans-directing ligands. Cyanide ion like carbon monoxide, being a -bonding ligand is high in the trans effect series because it tends to stabilize the transition state for reaction. The concept of - bonding in metal complexes was first introduced by Linus Pauling in order to account for the large stability of the cyanide complexes of transition metals as compared to the non-transition metals.

4. The cyanide ion, although it is without empty π orbitals of low energy, does have empty π^* Orbitals of only slightly higher energy. The π^* orbitals are of suitable symmetry to combine with the t_{2g} (π^*) of the metal.

   Electrons in the t_{2g} which are released towards them are said to undergo π - back-donation to the ligand. The ligand itself is known as a π -acceptor ligand. Clearly the effect of π -back-donation is to increase the covalent character of the bond between the metal and the coordinating atom, whereby the bond is also strengthened. This indicates that the metal-CN π- bonding is of importance in the stability of cyanide complexes and there is evidence of various types to support this. From close analysis of the vibrational spectra of cyanide complexes, the existence of -bonding has been confirmed more directly.

5. The cyanide ion does have the ability to stabilize the metal ions in low formal oxidation states as in the case of +2 oxidation state of iron obtaining in hemoglobin and myoglobin. It presumably does this by accepting electron density into its orbitals.

Among the other factors which influence the stability of complexes are:

I. Presence of chelate or multi-dentate ligands, rather than uni-dentate ones,
II. The size of the chelate ring,
III. Steric factors, and
IV. Resonance effects.

The increase in stability with dentate character is referred to as the chelate effect. Chelate compounds are even more stable when they contain a system of alternate double and single bonds. The -electron density is delocalized and spread over the ring, which is said to be stabilized by resonance. The best example of this includes porphyrin complexes with metals and haemoglobin in the RBC's is an iron-porphyrin complex.

The cytochromes are iron-containing haemoproteins in which the iron atom oscillates between Fe^{3+} and Fe^{2+} during oxidation and reduction. Several identifiable cytochromes occur in the respiratory chain, i.e; cytochromes b, c, and a. Depending upon the ligands present, the redox potential of a given cytrochome can be tailored to meet the specific need in the electron transfer scheme whether in photosynthesis or respiration. The potentials are such that the electron flow is b → c → a → O_2. At least some of the *a* type are thought to be capable of binding oxygen and reducing it. They are thus normally 5-coordinate in contrast to cytochrome c. They account for the unusual toxicity of the cyanide ion CN^- The latter binds tightly to the sixth position and stabilizes the Fe (III) to such an extent that it can no longer be readily reduced and take part in the electron shuttle. It is of some interest that cyanide, like the *iso*-electronic carbon monoxide molecule, binds to haemoglobin, but the inhibition of cytochrome a by the cyanide ion is much more serious than the interference with the oxygen transport. In fact, the standard treatment for cyanide poisoning is inhalation of amyl nitrite or injection of sodium nitrite to oxidize some of the haemoglobin to methemoglobin. The latter, although useless for oxygen transport binds cyanide even more tightly than haemoglobin or cytochrome a and removes it from the system.

CHAPTER 6

Oxidative Enzymes

1. Oxidative Enzymes

The examples of oxidative enzymes are Catalases and Peroxidases. They are iron-porphyrin proteins closely related in structure to haemoglobin and cytochromes. Both enzymes catalyse oxidation-reduction reactions in which H_2O_2 is the oxidizing agent, being reduced to water as some organic compound is oxidized. One molecule of catalase can decompose 44,000 molecules of peroxide per second.

Catalase activity is present in nearly all animal cells and organs; liver, erythrocytes, and kidney are rich sources. This activity has also been found in all plant materials which have been examined and in all micro-organisms other than obligate anaerobes. Peroxidases, however, are relatively rare in the animal world. All higher plants, on the other hand, are rich in peroxidase activity. Horse-radish is perhaps the richest source and has been extensively employed in the investigation of peroxidases.

The distinction between two types of enzymes is written as:

$$\begin{matrix} HO \\ HO \end{matrix} + \begin{matrix} HO \\ HO \end{matrix} \longrightarrow 2H_2O + O_2 \qquad (6.1)$$

Catalactic Reaction

$$\begin{matrix} HO \\ HO \end{matrix} + \begin{matrix} HO \\ HO \end{matrix} \quad R \longrightarrow 2H_2O + \begin{matrix} O \\ R \\ O \end{matrix} \qquad (6.2)$$

Peroxidatic Reaction

In this sense, the catalectic splitting of H_2O to water and O_2 becomes merely a special case of a peoxidatic reaction in which H_2O_2 serves both as a substrate and as acceptor.

Theorell has proposed that both classes of enzymes be given the common name hydroperoxidases indicating that their common substrate is hydrogen peroxide.

The **non-heme** iron proteins contain strongly bound functional iron atoms but no porphyrins. The iron atoms are bound by sulfur atoms. These proteins all participate in electron-transfer sequences.

Rubredoxins participate in a number of biological redox reactions, especially in anaerobic bacteria. Rubredoxin contains one iron atom tetrahedrally surrounded by four sulfur atoms, though the low resolution of the X-ray studies does not eliminate the possibility of some flattening of the tetrahedron.

Ferredoxins are relatively small proteins which contain sulfur-bound iron atoms, and like rubredoxin, participate in electron-transfer chains. Indeed, rubredoxin might logically be considered as one type of ferredoxin. The other types, to which the name is conventionally applied, contain two, four, or eight atoms of iron per molecule.

Iron storage and/or transport proteins are the body's method of handling these requirements. Ferritin and hemosiderin are iron storage proteins found in the liver, spleen, and bone marrow. Ferritin is a water-soluble, crystalline iron protein built up from apoferritin and micelles of a colloidal ferric-hydroxide-phosphate complex. Although the iron in ferritin is stored in the Fe^{3+} form it is incorporated and released in the Fe^{2+} form. Hemosiderin, on the other hand, is water soluble and is considered by some to be a dehydrated ferritin.

The major iron transport protein of blood plasma is a glycoprotein known as transferrin. It binds two atoms of iron per molecule so tightly that, for all practical purposes, there is no free plasma iron. Transferring releases iron to the red cell precursor by attaching to a receptor on the surface of the developing RBC. In microorganisms, iron is transported by substances called ferrichromes and ferrioxamines.

A person deficient in iron will become anaemic. Anaemias can be caused by excessive blood loss or decreased blood formation. Excessive blood loss can be caused by haemorrhages, menstruation, and bleeding ulcer. Iron-deficient anaemia is wide-spread throughout the world and is more common in women than in men. In particular, the pregnant or menstruating female must absorb greater percentage of her dietary iron.

II. Nitrogen Fixation

An enzyme system of particular importance is that which promotes the fixation of atmospheric nitrogen. This is of considerable interest for a variety of reasons. It is a very important step in the nitrogen cycle, providing available nitrogen for plant nutrition. It is an intriguing process since it occurs readily in various bacteria, blue-green algae, yeasts, and in symbiotic bacteria—legume associations under mild conditions.

Molecular nitrogen, N_2, is so unresponsive to ordinary chemical reactions that it has been characterized as "almost as inert as a noble gas." The very large triple bond energy (945 k J mol^{-1}) tends to make the activation energy prohibitively large. Thus, in spite of the fact that the overall enthalpy of formation of ammonia is exothermic by about 50 k J mol^{-1}, the common Haber process requires about 20 MPa pressure and 500°C temperature to proceed, even in the presence of the best Haber catalyst. In addition, to the purely pragmatic task of furnishing the huge supply of nitrogen compounds necessary for industrial and agricultural uses as cheaply as possible, the chemist is intrigued by the possibility of discovering processes that will work under less drastic conditions. We know they exist: We can *watch* a clover plant growing at 100k Pa and 25°C!

Nitrogen fixation may be studied under two heads:

In vivo **nitrogen fixation**

In vitro **nitrogen fixation**

In vivo **nitrogen fixation**

There are several bacteria and blue-green algae that can fix molecular nitrogen *in vivo*. Both free-living species and symbiotic species are involved. There are the strictly anaerobic *Clostridium pasteurianum*, facultative aerobes like *Klebsiella pneumoniae*, and strict aerobes aerobes like *Azotobacter vinelandii*. Even in the aerobic forms it appears that the nitrogen fixation takes place under essentially anaerobic conditions. The most important nitrogen-fixing species are the mutualistic species of *Rhizobium* living in root nodules of various species of legumes (clover, alfalfa, beans, peas, etc.).

The active enzyme in nitrogen fixation is *nitrogenase*. It is not a unique enzyme but appears to differ somewhat from species to species. Nevertheless the various enzymes are very similar. Two proteins are involved. The smaller has a molecular weight of 57000–73000. It contains an $Fe_4 S_4$ cluster. The larger protein is an $ά_2 B_2$ tetramer with a molecular weight of 220,000–240,000 containing two molybdenum atoms, about 30 iron atoms, and about 30 labile sulfide ions. The iron-sulfur clusters probably act as redox centres.

In Vitro Nitrogen Fixation

The discovery that molecular nitrogen was capable of forming stable complexes with transition metals led to extensive investigation of the possibility of fixation of nitrogen via such complexes. Of the various systems investigated, that employing titanium (II) was the first to be successful. Titanium (II) alkoxides form dinitrogen complexes which may then be reduced with subsequent release of ammonia or hydrazine. Such a process is not apt to be commercially competitive with the Haber process for the synthesis of ammonia but promises to be useful in the synthesis of other nitrogen compounds such as hydrazine and organic nitrogen compounds.

III. Biochemistry of Copper

Like iron, copper is indispensable for normal metabolism in man and most animals. Unlike iron, it is currently believed that most of the population obtains adequate amounts of copper from food, water, and cooking utensils. Thus copper supplements are probably not necessary.

Copper is solubilized in the acidic stomach and is absorbed from the stomach and upper small intestine. It is not known if the body regulates the absorption of copper, but current evidence indicates that regulation is accomplished by excretion of excess copper. Of the ~5mg of copper in the daily diet, 30% is absorbed. Of the amount absorbed, 80% is excreted through the bile, 16% is emptied directly back into the intestine through the gut walls, and only 4% is excreted in the urine. From the intestine, copper moves into the blood serum where it exists first as a copper-albumin complex. It then goes to the liver where copper becomes part of the copper protein ceruloplasmin. At equilibrium, 93% of serum copper is in ceruloplasmin and 7% in albumin- and amino acid-bound fractions. Ceruloplasmin copper is not released until the protein is catabolized. The albumin copper is probably the means used to transport copper to the liver, RBC, bone marrow, kidneys, and tissues. It appears that amino acid-bound copper is the form of copper for movement across membranes.

In the liver, copper is either, stored, incorporated into ceruloplasmin, or excreted in the bile. 60% copper in the RBC is found in the copper protein erythrocuprein and 40% as a labile non-erythrocuprein fraction. The level of copper in the kidneys is higher than one would predict from the small amount excreted in the urine.

The metabolic pathways of copper are sketched in Fig. III.

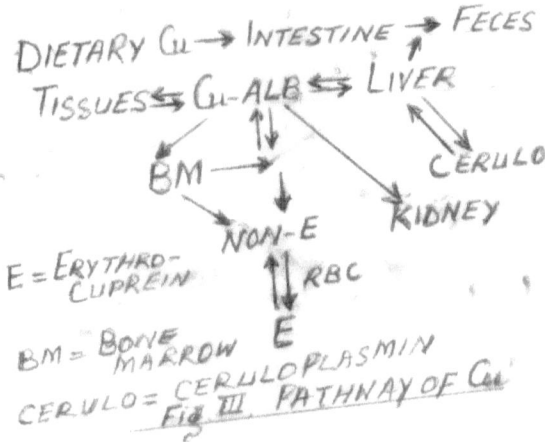

Several roles in body metabolism have been attributed to copper. One of these has to do with haemoglobin formation. Despite an adequate supply of iron, copper is required to prevent anaemic conditions. There have been three roles postulated for copper: Copper could facilitate iron absorption; Copper could be stimulatory to the enzymes in the heme / or globin biosynthetic pathways; and Copper could be involved in mobilization of stored iron, preparative to the incorporation of iron into the haemoglobin molecule.

Data concerning the role of copper in iron absorption are conflicting. A recent postulate is that ceruplasmin is really a ferrioxidase enzyme. It oxidizes ferrous iron to ferric for binding by transferrin.

Copper also is important in oxidative phosphorylation (ATP production by cellular respiration). Copper is a constituent of cytochrome oxidase, the terminal oxidase in the electron transport mechanism from which high-energy phosphate bonds are derived.

Copper is associated with the formation of aortic elastin. It may be that copper is necessary for amine oxidase activity and may play a role in the formation of cross linkages of elastin. Copper deficiency in pigs can result in a weakening of the aorta and other blood vessels. Copper is also a component of tyrosinase, an enzyme responsible for conversion of tyrosine to the black pigment melanin. A copper deficiency in animals may cause loss of hair colour which can be attributed to reduced tyrosinase activity. Albinism is associated with either an absence of or an inactive form of tyrosinase.

Human deficiencies have been identified in severely malnourished infants with chronic diarrhoea particularly those infants using modified cow's milk preparations that contain very little copper. More common, although still rare, is Wilson's disease, a condition of excess copper storage. Wilson's disease is of genetic origin. Patients have increased copper levels in liver, brain, kidney, and cornea. The symptoms are hepatic cirrhosis, brain damage, and kidney defects. They can be reversed by placing the patient on negative copper balance. This is usually done by diet and by use of the chelating agent pennicillamine.

The association of copper with body metabolism is shown in **Table IX**:

Table IX. Copper & Body Metabolism

Total Amount in Human Body = 2mg/Kg RDA = 900µ g	
Best Food Sources:	Liver, Kidney, Egg-yolk, Whole grains.
Absorption & Metabolism:	Limited Absorption, Transport by ceruloplasmin; stored in liver, excretion via bile.
Principal metabolic functions:	Formation of Hb (increases iron utilization), Constituent of oxidase enzymes.
Clinical manifestations of deficiency:	Hypochromic Anaemia, Excessive hepatic storage in Wilson's disease.

CHAPTER 7
Trace Metal Requirements in Plants and Animals

Zinc

The zinc ion is widely distributed in the body. Biochemically, it is associated with certain metalloenzymes. These include alcohol dehydrogenase, lactic dehyrogenase, malate dehyrogenase, D-lactate cytochrome c reductase, glyceraldehydes-3- phosphate dehyrogenase, glutamic dehyrogenase, aldolase, carbonic anhydrase, alkaline phosphatase, carboxypeptidase, and neutral protease. Zinc is also bound to RNA, stabilizing the secondary and tertiary structures.

Zinc is an essential dietary mineral, but currently it is felt that there is little justification in including it in mineral supplements. Foods rich in zinc include meat, milk, fish, nuts, and legumes.

Zinc deficiency is associated with impaired growth, parakeratosis (a thickened, scaly, inflamed skin), and retarded sexual maturation. Low plasma zinc levels are found in alcoholic cirrhosis, uremia, growth-retarded Iranian villagers, pregnancy, and in women taking oral contraceptives.

Zinc toxicity has resulted from the ingestion of acid food kept in a galvanized metal container and from industrial workers inhaling zinc oxide. Typical symptoms include chills, fever, malaise, coughing, salivation, and headache. Chelation therapy with dimercaprol promotes dramatic clinical improvement and a fall in blood zinc levels. The use of oral zinc sulphate has been suggested for wound healing. Currently, zinc sulphate is used as a typical astringent.

Chromium

Chromium is probably an essential element. It can be shown that chromium is necessary for optimal growth of experimental animals. In large quantities it is toxic. Chromium levels are higher in infants than adults. Chromium, however, seems to play some role in glucose tolerance.

The pharmacodynamic actions of chromium salts, chromates, and dichromates are very similar. They are destructive to tissue, regardless of whether applied topically or administered orally. When taken internally, they produce a characteristic nephritis and glycosuria. Persons exposed to "chromate dust" develop deep ulcers of the skin and nasal mucosa that heal very slowly.

Manganese

The total manganese content of an adult has been estimated to be 10 to 20 mg with the highest concentration occurring in bone and liver as well as in the pituitary, pineal, and lactating mammary glands. Manganese functions in many metallo-proteins as a non-specific cation. It is also associated with RNA and play a role in protein synthesis, oxidative phosphorylation, fatty acid metabolism, and cholesterol synthesis.

No deficiency state has been found in humans. The minimum daily requirements have been estimated at 3-9 mg. In other mammals, manganese deficiency is characterized by defective growth, bone abnormalities, reproductive dysfunction, central nervous system malfunctions. Excessive manganese intake can lead to chronic *manganism* (manganese poisoning). It is found mostly in the mining villages of Chile. Symptoms include mental disturbances, paresis (incomplete paralysis), and disturbances of gait. Chest hair will usually contain quite a high level of manganese. Manganese salts were once used as tonics but there is no known current therapeutic rationale for the administration of manganese.

Molybdenum

The requirements for molybdenum in human diet have not been established. It is present in all plant and animal tissues. The largest amounts (up to 3ppm) are found in liver, kidney, bone, and skin. Molybdenum has been found associated with flavin-dependent enzymes. The only current use of molybdenum today is as the oxide which together with ferrous sulphate in a specially co- precipitated complex is marketed as a hematinic preparation.

Lithium

The lithium ion is a depressant to the central nervous system and to circulation. The ion also has a diuretic action. Lithium is readily absorbed from the intestine and accumulates in the body. The extent of lithium accumulation is dependent upon the sodium intake.

Lithium salts have been advocated at different times as central nervous system depressants. The effectiveness of lithium carbonate in manic reactions is well documented. It corrects the patient's mood without the mental confusion that is usually observed following electro-convulsive therapy. Since lithium is toxic, serum lithium levels should be monitored.

Gold

Gold is toxic. It is slowly excreted by the kidney and will accumulate in the body. Regular determinations of plasma gold levels must be run for the patient on gold therapy. Gold toxicity involves the skin and mucus membranes, joints, blood, kidney, liver, and nervous tissue. If gold toxicity is severe, dimercaprol can be used to remove the accumulated gold from the body. Gold is used primarily in the treatment of rheumatoid arthritis.

CHAPTER 8

Metalloenzymes

I. Metalloenzymes

The outstanding characteristic feature of nearly all biochemical reactions is that they occur with great rapidity through the mediation of natural catalysts called enzymes.

Enzymes are large protein molecules. Not all proteins are catalysts, however. The functional groups in the composition of enzymes display properties that are not characteristic of them in low molecular compounds.

Metals play roles in approximately one-third of the enzymes. Metals may be a co-factor or they may be incorporated into the molecule, and these are known as **metalloenzymes.** Amino acids in peptide linkage possess groups that can form coordinate-covalent bonds with the metal atom. The free amino and carboxy group bind to the metal affecting the enzymes structure resulting in its active conformation.

Metals main function is to serve in electron transfer. Many enzymes can serve as electrophiles and some can serve as nucleophilic groups. This versatility explains metals frequent occurrence in enzymes. At least 50 metalloenzymes have been identified. The following metals have been found in metalloenzymes: Ca, Mn, Fe, Cu, Zn, and Mo. The most commonly used metals are Zn, Fe, and Cu.

Some metalloenzymes are listed below:

Fe: Aconitase, Phenylamine hydroxylase, Peroxidase, Catalase.

Cu: Laccase, Tyrosinase, Uricase.

Zn: Carbonic anhydrase, Carboxipeptidase.

Some metalloenzymes include hemoglobins, cytochromes, phosphotransferases, alcohol dehydogenase, arginase, ferredoxin, and cytochrome oxidase.

Metalloenzymes can be regulated in several ways since they are such a diverse group. One way metalloenzymes are regulated is the pH level. The pH level can disrupt the electron flow that the metal would normally help facilitate. In this way the pH level could inhibit the overall effectiveness of the metalloenzyme. Transition state analogues play a key role in the competitive inhibition of metalloenzymes because they mimic the structure of the substrate's transition state in the reaction of enzyme and the substrate. Metalloenzymes are incredibly diverse and function in a multitude of important physiological processes.

Structure of Metalloenzymes

Metalloenzymes are proteins which function as an enzyme and contain metals that are tightly bound and always isolated with the protein. In proteins such as haemoglobins and cytochromes, the metal is Fe^{2+} or Fe^{3+}, and it is part of the heme prosthetic group. In other metalloenzymes the metal is built into the structure of the enzyme molecule. The metal ion cannot be removed without destroying the structure of the enzyme. Metals built into the molecule include: most phosphotransferases, containing Mg^{2+}; alcohol dehydrogenase, Zn^{2+}; arginase, Mn^{2+}; ferredoxin, Fe^{2+}; and cytochrome oxidase, Cu^{2+}.

Metals are usually found in the active site of the enzyme. The metals resemble protons (H^+) in that they are able to accept an electron pair to form a chemical bond. In this respect, metals may act as general acids to react with anionic and neutral ligands. Metal's larger size relative to protons is compensated for their ability to react with more than one ligand. Metals typically react with two, four, or six ligands. A ligand is whatever molecule the metal interacts with. If a metal is bound with two ligands it will form a linear complex. If the metal reacts with four ligands the metal will be set in the centre of a square that is planar or it will form a tetrahedral structure. When six ligands react, the metal sits in the centre of an octahedron.

Amino acids in their peptide linkage in proteins possess groups with the ability to bind the metal resulting in coordinate-covalent bonds. The free amino and carboxyl groups in a protein can bind to the metal and this may bind the protein to a specific, active conformation. The fact that metals bind to several ligands is important in that metals play a role in bringing remote parts of the amino acid sequence together and help establish an active conformation of the enzyme.

Enzymes have a high degree of specificity and a tremendous efficiency; these permit *in vivo* reactions to occur rapidly through well-defined pathways.

They typically cause rates to increase 10^6 times or more compared to the un-catalyzed rate, or even to the rate attained with conventional non-enzyme catalysts. The specificity of enzymes can be classified as follows:

Absolute specificity, 2. Absolute group specificity, 3. Relative group specificity, and 4. Stereo-chemical specificity.

The specificity of an enzyme has to deal with its structure. The structure of an enzyme can be classified into four parts:

Primary structure, 2. Secondary structure 3. Tertiary structure and 4. Quaternary structure.

The catalytic action of enzymes is connected with the presence in the protein macro-molecules of certain sections playing the part of active centres. The active centre may be a part of the protein molecule or it may be a non-protein molecule. When it is a part of the protein molecule, it is called the recognition site; when it is a non-protein molecule, it is known as the prosthetic group. All metalloenzymes have metal ion as the prosthetic group.

Reactivity of Metalloenzymes

The incorporation of metal ions into enzyme structures can assist in the maintaining of a definite geometrical relationship between ionic and polar groups, through the geometric requirements of the coordinate bonds of the metal ion. Certain metal ions may also participate in the catalytic properties of the enzymes through ionic and coordinate bonding between the metal ion and the electron-donating groups of the enzyme and substrate, and through the ability of the metal ion to initiate redox reactions. Most of the metal ions that have biological functions have a co-ordination number of 6, with the donor groups arranged in an octahedral fashion. There are a few metals such as Mg^{2+} and Zn^{2+} that frequently coordinate only four donor groups tetrahedrally, and Cu^{2+} which has four coordinations to the corners of a square plane with the metal ion at the centre of the plane.

There is a saturation effect in the coordination of a metal ion by donor groups of both the enzyme and the substrate. Therefore, one would expect that the interaction of a free metal ion with the substrate would be greater than that of the metalloenzyme (in which the metal is already partially coordinated). If this were true, the metal ion would have a greater catalytic effect than the metalloenzyme. But reverse is always the case. This high activity of the enzyme is ascribed to the special environment of the substrate around the active site of the enzyme, through which additional binding of the substrate by adjacent organic groups of the enzyme takes place.

The activity of the enzymes greatly depends on the pH of the medium which a bio-chemical reaction proceeds in. the activity of some enzymes is many times superior to that of the known homogeneous and heterogeneous catalysts. Notwithstanding the great variety of kinds of chemical transformations occurring in the presence of the enzymes, the rates of reactions with their participation vary within rather narrow limits. Enzymes lower the activation energy of a chemical reaction.

To conduct certain enzyme-catalysed reactions, the presence of another substance called a co-enzyme is necessary in addition to the enzyme and the substrate. Co-enzymes undergo cyclic transformations in the course of these reactions. Vitamin B_6 and B_{12} are the well-known co-enzymes. The co-enzymes or co-factors generally act as acceptors or donors of a grouping or of atoms which are removed from or contributed to the substrate. The organic co-factors undoubtedly function in protein combination, and may be regarded as easily dissociable moieties of conjugated proteins. The type of co-enzyme concerned in the enzymatic process aids in classification. Since co-enzymes do not play the part of catalysts, but serve as a substrate, the name specialized substrates was recommended for them in recent years. In the course of a reaction, a specialized substrate transforms into a new product. As a result of the summary metabolic poly-enzyme process, no irreversible consumption of the specialized substrate occurs, and it is regenerated at the expense of other enzyme reactions in which the product it transforms into is used as a reactant. Thus the function of the specialized substrate consists in transferring the reactive intermediate from one enzyme process to another.

Ions of metals, besides forming metallo-enzymes, play a definite part in the catalytic properties of enzymes. They may be:

1. Inhibitors e.g; the ions Ag^+, Hg^+, and Pb^{2+};
2. Activators—Not directly participating in an act of catalysis, but facilitating the sorption of the substrate or the reaction of the substrate with an active centre, or the formation of a quaternary structure of the enzyme. It is reasonably certain that these activators also function in combination with the protein. Arginase, certain phosphatases, and some peptidases are examples of enzymes which require the presence of certain metal ions for their activity.

Experiments show that the mechanism of enzyme catalysis consists in processes of nucleophilic and electrophilic or general acid-base catalysis. Transferases, hydrolases, isomerases, lyases, and the majority of oxido-reductases were found to be acid-base catalysts. Reactions catalyzed by enzymes can be divided into two classes:

Reactions consisting in the transfer of electrons, and reactions attended by the transfer of both electrons and protons The reactivity of the enzymes is explained on the basis of Entactic State Theory put forth by R.J.P. Williams and his group. According to this theory, the enzyme and substrate first form an enzyme-substrate-complex (activated state) which later forms the products. There is spectroscopic and thermodynamic evidence in support of this theory.

General Functions of Metalloenzymes

Hemoglobin

A four sub-unit molecule, containing an iron atom in each sub-unit, in which each sub-unit binds a single molecule of oxygen. Hemoglobin transports oxygen from the lungs to the capillaries of the tissue.

Cytochromes

Cytochromes are integral membrane proteins. Cytochromes contain iron which serves to carry electrons between two segments of the electron-transport chain. The iron is reversibly oxidizable and serves as the actual electron acceptor for the cytochrome.

Phosphotransferase

The Mg^{2+} atom serves again in electron transfer.

Alcohol Dehydrogenase

It is a zinc metalloenzyme with broad specificity. They oxidize a range of aliphatic and aromatic alcohols to their corresponding aldehydes and ketones using NAD^+ as a co-enzyme.

Arginase

The metal atom of Mn^{2+} is used in electron transfer.

Ferredoxin

An electron transferring protein involved in one-electron transfer processes.

Cytochrome Oxidase

The copper ions easily accommodate electron removed from a substrate and can just easily transfer it to a molecule of oxygen.

Carboxypeptidase A (CPA)

Carboxypeptidase A (CPA) is a Zinc metalloenzyme that breaks peptide linkages in the digestion of proteins. The Zinc ion that the enzyme contains in its active site plays a key role in that function. This metalloenzyme can be regulated by diet. The source of zinc in

humans is almost entirely through diet. The zinc atom serves as a metal ion catalyst and promotes hydrolysis. The substrate fits into the hydrophobic pocket in CPA and zinc binds to the carboxyl group of the substrate to help stabilize the enzyme-substrate complex. The zinc ion acts as a generalized acid and stabilizes the developing O^- as water attacks the carbonyl.

The enzyme occurs in the pancreas of many species ranging from dogfish to man. Two types of this enzyme are known—CPA and CPB. CPA is secreted in the pancreas as an inactive zymogen, procarboxypeptidase A (PCA). Four forms of CPA arise from enzymatic release from bovine PCA, and detailed crystallographic work has been carried out on one of these, CPAά. It has a molecular weight of 34, 472 and contains 307 amino acid residues and one zinc atom. The enzyme specifically catalyses the hydrolysis of the carboxy-terminal peptide bonds in protein and peptide substrates.

The peptide bond which is hydrolysed must be adjacent to a C-terminal free carboxylate ion and the rate of hydrolysis is enhanced if the side-chain of the C-terminal residue (R_1') is aromatic or branched aliphatic of L-configuration. Hydrolysis occurs at the position indicated by the dotted line:

$$\begin{array}{c} R_2 \\ | \\ CH-C-NH-C-C-N-CH-COO^- \\ \|H\| \\ OO \end{array}$$

with R_1 and R_1' on the appropriate carbons and H on the nitrogen.

Hydrolysis of Peptide Bond

Mechanism of Action

The structure of CPA was the first to be determined for a metalloenzyme. In CPAá the zinc ion is bound in a distorted tetrahedral arrangement, with two histidine-nitrogen atoms and one glutamate-carboxyl oxygen atom. Fourth coordination is completed by a water molecule. Once the peptide bond has been broken with formation of the acid-anhydride, rapid hydrolysis of the anhydride would occur. The products would then vacate the active site, leaving it ready to bind another molecule of substrate and repeat the cycle.

When the enzyme binds to a substrate a number of interactions develop and there are major conformational changes in the structure of the enzyme molecule. Chief effects of the binding of a substrate by the active site in CPAá are:

Displacement of the water molecule coordinated to zinc by the oxygen of the carbonyl group of the susceptible peptide link on the substrate; Formation of a bond between the negatively charged terminal carboxylate group of the substrate with the positively charged guanidinium group of arginine; Insertion of the COOH-terminal side-chain into the "pocket" of CPAá, thereby displacing several of the water molecules originally there; this region now becomes largely hydrophobic in character; Movement of the tyrosine residue of CPAá by a twisting of the carbon-carbon chain so that its OH group is now only 300 pm from the susceptible peptide link of the substrate; probably this residue acts as proton donor to the NH group beyond hydrolysis; Movement of glutamate; this is implicated in the hydrolysis reaction and may provide the base for the attack of the carbon atom of the susceptible peptide bond.

Regulation and Control of Metalloenzymes

Approximately one-third of the known enzymes have metals as part of their structure, require that metals be added for activity, or are further activated by metals. Because the grouping, metalloenzymes, is so large and broad it would be almost impossible to explain how all of them can be controlled and regulated. In light of this, it is important to instead mention how the important functions of metals in enzymes can be disrupted and thus inhibited.

Metals resemble protons (H^+) in that they are electrophiles that are capable of accepting an electron pair to form a chemical bond. In doing so, metals may act as general acids to react with anionic and neutral ligands. This characteristic of metals is helpful in enzymatic structure and function. Changes in pH can disrupt the electron flow that the metal would normally help facilitate and thus inhibit the overall effectiveness of the metalloenzyme. Also because of the variability inherent to the metal's ability to react with more than one ligand, we see metals as part of the active site in many metalloenzymes.

Metals play a large role in the activity of a multitude of biological molecules. Without the intake of metals such as zinc in the diet, one almost certainly inhibits the production and / or activity of many vital enzymes. Among enzymes that would not be produced by the body were it not for the presence of zinc in the body are carbonic anhydrase, the carboxypeptidases, alkaline phosphatase, lactic acid, and alcohol dehydrogenases. The recommended daily allowance for zinc is 15 mg. with 20 and 25 mg. during pregnancy and lactation. The average adult human being ingests 12 to 20 mg. of zinc per day. Deficiencies in dietary zinc intake can result in stunted growth, enlarged liver and spleen, and under development of genitals and secondary sex characteristics. Outside of dietary intake deficiencies of zinc, and thus in enzymes that contain zinc, can be caused by the excretion of zinc in perspiration, or by blood loss if there is parasite infection. There is also increasing evidence that zinc plays an important role in protein bio-synthesis and utilization. The addition of small amounts of zinc to a diet containing suboptimal amounts of a vegetable protein, as indicated by the growth of young rats, causes a pronounced increase in protein utilization and growth. This defect may result from a failure in adequate RNA synthesis. Zinc apparently inhibits the enzyme ribonuclease. Thus, in zinc deficiency, excessive destruction of RNA could occur. This demonstrates that the dietary intake of metal is not only important for the production of key enzymes but also for the inhibition of many others.

II. Vitamin B_{12}

Vitamin B_{12} is a member of the vitamin B complex. It contains cobalt, and so is also known as **cobalamin**. It is exclusively synthesized by bacteria and is found primarily in meat, eggs, and dairy products. The name vitamin B_{12} (or B_{12}, for short) is used in two different ways:

In a broad sense, it refers to group of cobalt-containing compounds known as cobalamins—**cyanocobalamin** (an artifact formed as a result of the use of cyanide in the purification procedures), **hydroxocobalmin** and the two co-enzyme forms of B_{12}, **methylcobalamin** (MeB$_{12}$) and **5-deoxyadenosylcobalamin** (adenosylcobalamin-AdoB$_{12}$).

In a more specific way, the term B_{12} is used to refer to only one of these forms, **cyanocobalmin**, which is the principal B_{12} form used for foods and in nutritional supplements.

Pseudo-B_{12} refers to B_{12}-like substances which are found in certain organisms, such as *Spirulina spp.*(blue-green algae, cyanobacteria). However, these substances do not have B_{12} biological activity for humans.

Structure

Vitamin B_{12} is a cobalt (III) complex which was first isolated in 1948 independently by Lester Smith in England and by Rickes *et. al.* in the United States. Subsequently, its crystal structure was determined by X-ray diffraction and its chemical properties intensively studied. Vitamin B_{12} was first isolated from the liver (it had been known for a long time before this vitamin was discovered that sufferers from pernicious anaemia showed marked improvement when fed on a diet of raw liver) and now is produced commercially by fermentation processes using micro-organisms such as species of *Streptomyces* and *Propionibacter* and also *Bacillus megatherium* and *Nocardia rugosa.*

In 1973, a team of organic chemists headed by Prof. R.B.Woodward at Harvard University and Prof. Albert Eschenmoser at the Swiss Fedral Institute of Technology completed a laboratory synthesis of vitamin B_{12}. The synthesis was a notable achievement. In terms of effort expended, it is undoubtedly the most impressive organic synthesis ever done. More than 90 separate synthetic reactions were required and dozens of highly trained chemists labored for years to bring it to completion. It remains one of the classic feats of total synthesis. B_{12} cannot be made by plants or animals, as the only type of organisms that have the requisite enzymes are bacteria and *archaea*.

The **molecular formula** of Cyanocobalamin is $C_{63} H_{88} Co N_{14} O_{14} P$

The **molecular mass** is 1355.37g/mol.

B_{12} is the most chemically complex of all the vitamins. The structure of B_{12} is based on a corrin ring, which, although similar to the porphyrin ring found in heme, chlorophyll, and cytochrome, has two of the pyrrole rings directly bonded. The central metal ion is Cobalt. Four of the six co-ordinations are provided by the corrin ring nitrogens, and a fifth by a *dimethylbenzimidazole* group. The sixth coordination partner varies, being a cyano group (–CN), a hydroxyl group (–OH), a methyl group (–CH$_3$) or a 5-deoxyadenosyl group (here the C5 atom of the deoxyribose forms the covalent bond with Co), respectively, to yield the four B$_{12}$ forms mentioned above. The covalent C—Co bond is the only carbon-metal bond known in biology.

Functions

The primary functions of Vitamin B$_{12}$ are:

1. It is important in the formation of RBC's
2. It is important in maintaining healthy nervous system.
3. It is especially important in tissues where cells are dividing rapidly.
4. B$_{12}$ is necessary for the rapid synthesis of DNA during cell division.
5. It plays a vital role in the metabolism of fatty acids essential for the maintenance of myelin.

Deficiency Symptoms

If B$_{12}$ deficiency occurs, DNA production is disrupted and abnormal cells called *megaloblasts* occur. This results in anemia. Symptoms include excessiveness tiredness, breathlessness, listlessness, pallor, and poor resistance to infection. Other symptoms include a smooth, sore tongue and menstrual disorders.

Prolonged B$_{12}$ deficiency can lead to nerve degeneration and irreversible neurological damage. When deficiency occurs, it is more commonly linked to a failure to effectively absorb B$_{12}$ from the intestine rather than dietary deficiency. Absorption of B$_{12}$ requires the secretion from the cells lining the stomach of a glycoprotein, known as intrinsic factor. The B$_{12}$ intrinsic factor complex is then absorbed in the ileum (part of small intestine) in the presence of calcium. Certain people are unable to produce intrinsic factor and the subsequent pernicious anaemia is treated with injections of B$_{12}$.

Dietary Sources

The only reliable unfortified sources of vitamin B$_{12}$ are meat, dairy products, and eggs. There has been considerable research into possible plant food sources of vitamin B$_{12}$. The current nutritional consensus is that no plant foods can be relied on as a safe source of vitamin B$_{12}$.

The Structure of Cyanocobaamin is given in fig.IV:

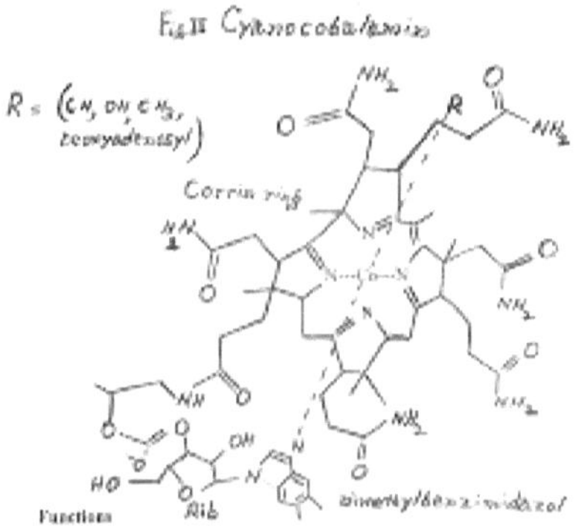

Fig II Cyanocobalamin

Human faeces can contain significant B_{12}. A study has shown that a group of Iranian vegans obtained adequate B_{12} from unwashed vegetables which had been fertilized with human manure. Fecal contamination of vegetables and other plant foods can make a significant contribution to dietary needs, particularly in areas where hygiene standards may be low. This may be responsible for the lack of anaemia due to B_{12} deficiency in vegan communities in developing countries.

Good sources of vitamin B_{12} for vegetarians are dairy products or free-range eggs. ½ pint of milk (full fat or semi-skimmed) contains 1.2µg. A slice of vegetarian cheddar cheese (40g) contains 0.5µg. A boiled egg contains 0.7µg. Fermentation in the manufacture of yoghurt destroys much of the B_{12} present. Boiling milk can also destroy much of the B_{12}µ.

Vegans are recommended to ensure their diet includes foods fortified with vitamin B_{12}. A range of B_{12} fortified foods are available. These include yeast extracts, Vecon- vegetable stock, Veggie-burger mixes, textured vegetable protein, Soya milks, vegetable and sunflower margarines, and breakfast cereals.

Required Intakes

The old Recommended Daily Amounts (RDA's) have now been replaced by the term Reference Nutrient Intake (RNI). The RNI is the amount of nutrient which is enough for at least 97% of the population.

RNI's for Vitamin B_{12} are given in **Table X** below:

Table X. RNI's for Vitamin B_{12}, µg/day. (1000µg = 1 mg)

AGE RNI
0 to 6 months 0.3 µ g
7 to 12 months 0.4 µ g
1 to 3 yrs. 0. 5 µ g 4 to 6 yrs. 0.8 µ g
7 to 10 yrs. 1.0 µ g
11to 14 yrs 1.2 µ g
Breast Feeding Women 2.0 µ g

Pregnant women are not thought to require any extra B_{12}, though little is known about this. Lactating women 15 + yrs. 1.5µ g need extra B_{12} to ensure an adequate supply in breast milk.

B_{12} has very low toxicity, and high intakes are not thought to be dangerous.

CHAPTER 9

Photosynthesis

The photoynthetic process in green plants consists of splitting the elements of water, followed by reduction of carbon dioxide:

$$2H_2O \longrightarrow [2H_2] + O_2 \qquad (9.1)$$

$$CO_2 + [2H_2] \longrightarrow 1/x\,(CH_2O)_x + H_2O \qquad (9.2)$$

Where $[2H_2]$ does not imply the hydrogen but the reducing capacity formed by the oxidation-reduction of water. The German surgeon, Julius Robert Mayer, who recognized that plants convert solar energy into chemical energy, said:

"Nature has put itself the problem of how to catch in flight light streaming to the Earth and to store the most elusive of all powers in rigid form. The plants take in one form of power, light; and produce another power, chemical difference."

The actual chemical equation which takes place is the reaction between carbon dioxide and water, catalyzed by sunlight, to produce *glucose* and a waste product, oxygen. The glucose sugar is either directly used as an energy source by the plant for metabolism or growth, or is polymerized to form *starch*, so it can be stored until needed.

The waste oxygen is excreted into the atmosphere, where it is made use of by plants and animals for respiration.

$$6CO_2 + 6H_2O \longrightarrow C_6H_{12}O_6 + 6\,O_2 \qquad (9.3)$$

In green plants there are two photosynthetic systems. The two differ in the type of chlorophyll present and in the accessory chemicals for processing the trapped energy of the photon. Chlorophyll is an important compound of a very important group of bio-inorganic compounds containing metals, the metalloporphyrins. Perhaps the most important class of metal-containing compounds in biological systems is that comprising complexes between metal ions and porphyrin ligands. The latter are macrocyclic tetrapyrrole systems with conjugated double bonds and various groups attached to the perimeter. Two metalloporphyrins of great biological importance are: Heme, which is an iron (II) porphyrin; and Chlorophyll, which is a magnesium complex containing a modified ring system.

Chlorophyll

Chlorophyll is the molecule that traps this "most elusive of all powers"—and is called a photoreceptor. It is found in the chloroplasts of green plants, and is what makes green plants, green. The basic structure of a chlorophyll molecule is a porphyrin ring, coordinated to a central atom. This is very similar in structure to the heme group found in haemoglobin, except that in heme group the central atom is iron, whereas in chlorophyll it is magnesium.

There are actually 2 types of chlorophyll, named *a* and *b*. They differ only slightly, in the composition of a side chain (in *a* it is $-CH_3$, in *b* it is CHO). Both of these two chlorophylls are very effective photoreceptors because they contain a network of alternating single and double bonds, and the orbitals can delocalize stabilizing the structure. Such delocalized polyenes have very strong absorption bands in the visible regions of the spectrum, allowing the plant to absorb the energy from sunlight.

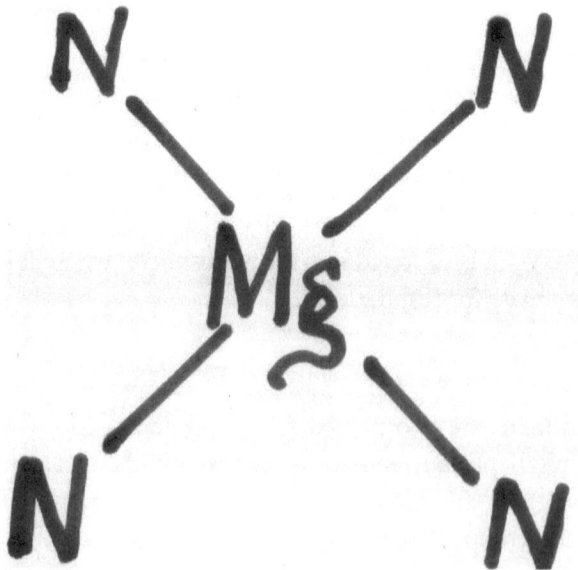

Fig IV. Central core of Chlorophyll.

Absorption Spectra of Chlorophyll a and b are given in Fig.V:

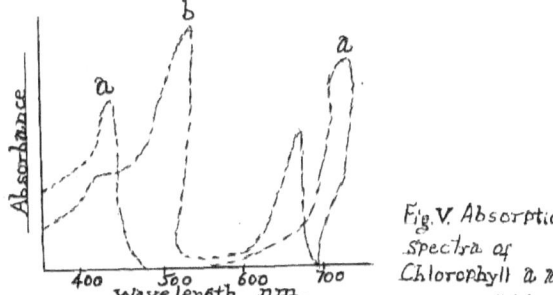

Fig.V. Absorption Spectra of Chlorophyll a and b

The different side-groups in the 2 chlorophylls "tune" the absorption spectrum to slightly different wavelengths, so that light that is not significantly absorbed by chlorophyll a, at, say, 460 nm, will instead be captured by chlorophyll b, which absorbs strongly at that wavelength. Thus these two kinds of chlorophyll complement each other in absorbing sunlight. Plants can obtain all their energy requirements from the blue and red parts of the spectrum, however, there is still a large spectral region, between 500-600nm, where very little light is absorbed. This light is in the green region of the spectrum, and since it is reflected, this is the reason plants appear green. Chlorophyll absorbs so strongly that it can mask other less intense colors. Some of these more delicate colors (from molecules such as carotene and quercetin) are revealed when the chlorophyll molecule decays in the autumn, and the woodlands turn red, orange, and golden brown. Chlorophyll can also be damaged when vegetation is cooked, since the central magnesium atom is replaced by hydrogen ions. This affects the energy levels within the molecule, causing its absorption spectrum to alter. Thus cooked leaves change color---often becoming a paler, insipid yellowy green. The chlorophyll molecule is the active part that absorbs the sunlight, but just as with hemoglobin, in order to do its job (synthesizing carbohydrates) it needs to be attached to the backbone of a very complicated protein. This protein may look haphazard in design, but it has exactly the correct structure to orient the chlorophyll molecules in the optimal position to enable them to react with nearby CO_2 and H_2O molecules in a very efficient manner. Several chlorophyll molecules are lurking inside this bacterial photoreceptor protein.

The different side-groups in the 2 chlorophylls "tune" the absorption spectrum to slightly different wavelengths, so that light that is not significantly absorbed by chlorophyll a, at, say, 460 nm, will instead be captured by chlorophyll b, which absorbs strongly at that wavelength. Thus these two kinds of chlorophyll complement each other in absorbing sunlight. Plants can obtain all their energy requirements from the blue and red parts of the spectrum, however, there is still a large spectral region, between 500–600nm, where very little light is absorbed. This light is in the green region of the spectrum, and since it is reflected, this is the reason plants appear green. Chlorophyll absorbs so strongly that it can mask other less intense colours. Some of these more delicate colours (from molecules such as carotene and quercetin) are revealed when the chlorophyll molecule decays in the autumn, and the woodlands turn red, orange, and golden brown.

Chlorophyll can also be damaged when vegetation is cooked, since the central magnesium atom is replaced by hydrogen ions. This affects the energy levels within the molecule, causing its absorption spectrum to alter. Thus cooked leaves change colour—often becoming a paler, insipid yellowy green. The chlorophyll molecule is the active part that absorbs the sunlight, but just as with haemoglobin, in order to do its job (synthesizing carbohydrates) it needs to be attached to the backbone of a very complicated protein.

This protein may look haphazard in design, but it has exactly the correct structure to orient the chlorophyll molecules in the optimal position to enable them to react with nearby CO_2 and H_2O molecules in a very efficient manner. Several chlorophyll molecules are lurking inside this bacterial photoreceptor protein.

ADP and ATP
Adenosine triphosphate (ATP) is formed to a certain extent directly in photosynthesis and is the end product of respiration. It may then be used as an energy source to drive the many chemical reactions of the cell. The energy of ATP comes from the hydrolysis of a polyphosphate linkage. The reaction is usually represented as:

Magnesium is necessary for the conversion to be effected and Mg^{2+} complexes are involved. The standard free energy change for the hydrolysis of ATP to ADP is estimated to be -31.0 kJ mol^{-1}. Because of concentrations obtaining in the cell, the available free energy for bio-synthesis is thought to be about -40 to -50 kJ mol^{-1}

CHAPTER 10
Biochemistry of Nonmetals

I. Biochemistry of the Nonmetals

Many of the non-metals such as hydrogen, carbon, nitrogen, oxygen, phosphorus, sulfur, chlorine, and iodine are essential elements, and most are used in quantities far beyond the trace levels. Nevertheless, most of the chemistry of these elements in biological systems is more closely associated with organic chemistry than wit inorganic chemistry.

Structural uses

There are three important minerals used by organisms to form hard tissues such as bones and shells. These are: 1. **Calcium carbonate**, 2. **Silica**, and 3. **Apatite**. The most widespread of these is calcium carbonate, an important structural component in animals ranging from Protozoa to Mollusca and Echinodermata. It is also a minor component of vertebrate bones. Its widespread use is probably related to the generally uniform distribution of dissolved calcium carbonate.

Animals employing calcium carbonate are most abundant in fresh waters containing large amounts of calcium and magnesium ("hard water") and in warm, shallow seas where the partial pressure of carbon dioxide is low (e.g; the formation of coral reefs by coelenterates). The successful precipitation of calcium carbonate depends upon the equilibrium:

$$Ca^{2+} + 2\ HCO_3 \rightarrow Ca\ CO_3 + CO_2 + H_2O \qquad (10.1)$$

And is favoured by high $[Ca^{2+}]$ and low $[CO_2]$. Nevertheless, organisms exhibit a remarkable ability to deposit calcium carbonate from hostile environments. A few fresh water clams and snails are able to build reasonably large and thick shells in lakes with a pH of 5.7–6.0 and as little as 1.1 ppm dissolved calcium carbonate.

It is of interest that two thermodynamically unstable forms of calcium carbonate, aragonite and vaterite, are found in living organisms as well as the more stable calcite. There appears to be no simple explanation for the distribution of the different forms in the various species.

Silica is found in the tissues of the primitive algal phyla Pyrrhophyta (dinoflagellates) and Chrysophyta (diatoms and silicoflagellates). One family of higher plants, the Equisetaceae, or horsetails, contains gritty deposits of silica—hence their name "scouring rushes".

Some protozoa (radiolarians), Gastropoda (limpets), and Porifera (glass sponges) employ silica as a structural component. Silicon is an essential element in chicks and rats and is probably necessary for proper bone growth in all higher animals.

The third type of compound used extensively as a structural component is apatite, $Ca_5(PO_4)_3 X$. Hydroxyapatite (X = OH) is the major component of bone tissue in the vertebrate skeleton. It is also the principal strengthening material in teeth. Partial formation of Fluorapatite (X = F) strengthens the structure and causes it to be less soluble in the acid formed from fermenting organic material. Application of stannous fluoride to the teeth, especially to carious enamel, probably results in the conversion of the hydroxyapatite to a new material. In the laboratory, at least two compounds have been characterized; $Sn_2(OH) PO_4$ formed at low concentrations of SnF_2, and $Sn_3 F_3 PO_4$ formed at higher concentrations. Presumably these conversions of hydroxyapatite are involved in the reduction of caries by fluoride ions and stannous fluoride. Fluorapatite is also used structurally in certain Brachiopod shells.

II. Medicinal Chemistry

1. Antibiotics

The suggested antibiotic action of transferring is typical of the possible action of several antibiotics in tying up essential metal ions. Streptomycin, aspergillic acid, usnic acid, the tetracyclines, and other antibiotics are known to have chelating properties. Presumably some antibiotics are delicately balanced so as to be able to compete successfully with the metal-binding agents of the bacteria while not disturbing the metal processing by the host. There is evidence that some bacteria have developed resistance to antibiotics through the development of altered enzyme systems that can compete successfully with the antibiotic. The action of the antibiotic need not be a simple competitive one. The chelating properties of the antibiotic may be used in metal transport across membranes or to attach the antibiotic to a specific site from which it can interfere with the growth of bacteria.

The behaviour of valinomycin is typical of a group known as "ionophore antibiotics". These compounds resemble the crown ethers and cryptates by having several oxygen or nitrogen atoms spaced along a chain or ring that can wrap around a metal ion.

These antibiotics are useless in man because they are toxic to mammalian cells, but some of them find use in treating coccidiosis in chickens. The toxicity arises from the ion-transporting ability.

The tetracyclines form an important group of antibiotics. Their activity appears to result from their ability to chelate metals since the extent of antibacterial activity parallels the ability to form stable chelates. The metal in question appears to be magnesium since addition of large amounts of magnesium can inhibit the antibiotic effects of the tetracycline. It is not certain how the tetracyclines coordinate magnesium, but the molecular structure of the potassium salt of oxytetracycline is known and shows excessive chelation.

2. Chelation Therapy

Chelation therapy is a conventional therapy used to treat heavy metal poisoning. However, it is also promoted as a type of alternate therapy to treat heart diseases, cancer, and other conditions. It most often involves the injection of EDTA (ethylene diamine tetra acetic acid), a chemical that binds (chelates) heavy metals which include iron, mercury, cadmium, and zinc. The term "chelation" comes from the Greek word "chele." Which means claw, referring to the way the chemical grabs on to these metals.

Chelation therapy has also been promoted as an alternative treatment for many unrelated conditions, such as gangrene, thyroid disorders, multiple sclerosis, muscular dystrophy, psoriasis, diabetes, arthritis, Alzheimer's disease, and the improvement of memory, sight, hearing, and smell.

Chelating agents can be used therapeutically to treat problems caused by the presence of toxic elements. We have seen that an essential element can be toxic if present in too great a quantity. This is the case in Wilson's disease (hepatolenticular degeneration), a genetic disease involving the build-up of excessive quantities of copper in the body. Many chelating agents have been used to remove the excess copper, but one of the best is D-penicillamine, $HSC(CH_3)_2CH(NH_2)COOH$. This chelating agent forms a complex with copper ions that is colored an intense purple and, surprisingly, has a molecular weight of 2600.

The body has essentially no means of eliminating iron, so an excessive intake of iron causes various problems known as siderosis. Chelating agents are used to treat the excessive build-up of iron. In many cases these chelates resemble the analogous compounds used by bacteria to chelate iron. The ideal chelating agent will be specific for the metal to be detoxified since a more general chelating agent is apt to cause problems by altering the balance of other essential metals. The concepts of hard and soft metal ions and ligands can be used to aid in this process of designing therapeutic chelators.

A slightly different mode of therapy involves the use of *cis*-diamminedichloroplatinum (II), $Pt(NH_3)_2Cl_2$.

The exact action of the drug is not known, but only the *cis*-isomer is active, not the *trans*-isomer.

There is no evidence the chelation therapy is a safe treatment for any type of cancer. Chelation therapy may produce toxic effects, including kidney damage, irregular heartbeat, and swelling of the veins. It may also cause nausea, vomiting, diarrhoea, and temporary lowering of blood pressure. Chelation therapy may be dangerous in people with kidney disease, liver disease, or bleeding disorders. Women who are pregnant or breast-feeding should not use this method.

CHAPTER 11

Problems in Ecosystem

It is unfortunate that most of the problems in our ecosystem that are potentially solvable by the application of inorganic chemistry are those dealing with the deleterious effects of inorganic chemicals. Thus we find that many of the major sources of concern are inorganic species such as mercury in water, sulfur dioxide and nitrogen oxides in the air, and lead practically everywhere. To a larger extent many of the key issues fall outside the scope of pure chemistry, even outside the scope of pure science, into the areas of medicine, law, sociology, and politics. Nevertheless, in this section a brief survey of the inorganic chemistry relevant to these problems will be given.

I. The Historical Perspective

Until man developed cultural traits such as language, the use of tools and fire, and agriculture, he was little different from the other predators that roamed the plains and forests. The development of agriculture resulted in modification of the ecosystem, but in a manner not unlike recurring natural phenomena. An abandoned field returns to the forest as rapidly as an area burnt over by a forest fire. The advent of civilization resulted in increased consumption, and irreversible changes began to occur. One example is the deforestation of the lands to provide timber for construction, a major component of commerce. Deforestation was followed by erosion and in some areas intensive grazing. Fundamental changes in the flora, fauna, soil, and even climate can take place under these circumstances.

Despite the fact that such stresses on the ecosystem are as old as recorded history, it was not until the Industrial Revolution that man had sufficient control of his environment to destroy it. This was relatively recent as man's history goes; yet the problem continued to develop for about 200 years before solutions were sought. Even today there is a lot of whistling-in- the- dark "theorizing" by some people, such as: "If flies can become resistant to DDT why can't man."

Since there appears to be little prospect that man is willing to reduce his population level to that which existed when he was a hunter and food-gatherer, and even less that he will voluntarily return to that earlier level of creature comforts, the problem facing us is how to maximize the availability of the products of civilization while minimizing the cost in terms of pollution, depletion of natural resources, and related problems.

II. Agriculture

In view of the overwhelming population problem there has been a tendency to seek simplistic solutions by increasing the food supply via the application of fertilizers and insecticides. The former have proved valuable in increasing crop yields but have been a source of environmental degradation via run-off and leaching. In this way large quantities of nitrogen, phosphate, and potassium enter the ecosystem resulting in (1) explosive growth of algae in rivers and lakes, and (2) stagnant sloughs as the excess algae die and rot without normal consumption and utilization in the food chain.

There is no panacea for the problem of pollution from agricultural fertilizers. It is apparent that more selectivity will have to be used in the future—dumping huge amounts of lime or a nitrate-phosphate-potassium fertilizer per square kilometer may not solve a nutritional problem resulting from a trace element deficiency. It may even worsen the problem since several metal ions are rendered unavailable by precipitation by phosphate or hydroxide:

$$3\ Mn^{2+} + 2PO_4^{3-} \rightarrow Mn_3(PO_4)_2 \qquad (11.1)$$

$$Fe^{n+} + nOH^- \rightarrow Fe(OH)_n\ n = 2, 3 \qquad (11.2)$$

There is a large and growing literature on the importance of trace elements in plant and animal nutrition, and the future should see increasing applications of the principle of selective nutritional adjustments. The use of "slow-release" fertilizers can also reduce the portion wasted in run-off.

The use of herbicides and insecticides has also posed a problem. Their exclusive use in control is seldom practical. The target develops resistance and even larger doses are applied. Sooner or later, non-target organisms become involved. Often these chemicals owe their usefulness to the inhibition of enzyme action and when concentrated by the food chain in non-target organisms, they can cause unforeseen effects.

In contrast to problems arising from the lack of essential elements, problems may also arise through the overabundance of certain elements. Obviously land may be poisoned and made unfit for agriculture through the natural presence or application of elements which are toxic in high concentrations. More subtle is the production of non-arable land simply through agricultural use, not via depletion but by accumulation. Continuous irrigation in arid lands leads to a build-up in the soil of ionic solutes as the water evaporates without sufficient rainfall to leach out these solutes. This type of poisoning is merely a matter of ionic strength and osmotic competition with the plants rather than the specific action of any particular element.

Prof. (Dr) R.D.Gupta, ex-Associate Dean cum Chief Scientist KVK SKAUST-J, in an article appearing in *Daily Excelsior* dated December 27, 2006 writes:

"'Green Revolution' based upon high yielding varieties of various crops, chemical fertilizers have no doubt helped in making India self-sufficient in food grain production. However, Agricultural Scientists have now been able to find out the injurious effects of chemical fertilizers and pesticides on the soils. Use of fertilizers has, in fact, depleted the soils of their nutrients, microflora and microfauna which has led to serious reduction in the crop yields. Ludhiana District (Punjab) although has recorded highest yields of a number of crops yet it has shown the maximum deficiencies of plant micro- nutrients."

"Apart from using N, P, and K_2O fertilizers, now many farmers of Panjab have commenced to use $ZnSO_4$ as zinc fertilizer to coup with the growing soil Zn depletion which has evinced to reduce the yields. It is because Zn is one among many several micronutrients essential for plant growth like micronutrients Viz; C, H, O, N, P, K, ca, Mg, and S. The other six micronutrients are: Fe, Cu, Mn, B, and Cl. The latest studies have shown great impact of micronutrients on high-yielding varieties of crops. For example, a rice-wheat cropping sequence yielding 88 ha^{-1} year^{-1} removes 663 Kg N, P, and K, and several Kg micronutrients causing thereby a serious drain on the plant-nutrient reserve in the soil. It therefore, becomes necessary for the farmers to put N,P, K back into the soil to get higher yield of crops, vegetables and fruits. Most of the farmers, however, add only nitrogenous fertilizers (urea, ammonium sulfate) as other fertilizers are beyond their means. Thus, it is not surprising that P, K, and micronutrient deficiency has become severe in the intensively cropped areas. The deficiency of P, and K has been substantiated by the National Average N : P_2O_5 : K_2O ratio of 8:3: 3:1 against 4:2:1."

Thus indiscriminate use of chemical fertilizers has created deficiency both of macro (Pand K) and micronutrients. A study conducted at Ranchi Agriculturtal College farm during 2004–5 showed that use of 100Kg NPK (50, 25, 25) has led to the depletion of Zn by 629 gha^{-1}, Cu by 433 gha^{-1}, Fe by 41809 ha^{-1} and Mn by 4185 ha^{-1}.

Their depletion was directly proportional to NPK use. In the years to come deficiency of Zn would further magnify and that of other micronutrients like Fe, Mn, and Cu would crop up if inherently poor soils are continuously exploited even at this level of production. According to ICAR reports in areas where Zn deficiency was reduced, Fe and other micro-nutrient deficiency started to arise."

"Application of urea alone has also made many of the soils acidic and contaminated the drinking water with NO_3. NO_3 contaminated water has produced "blue baby disease" in many parts of India among the babies. Nitric oxides (NO, N_2O, NO_2 etc.) are being increased in the atmosphere. These are one of the agents responsible for depleting Ozone layer."

"Use of chemical fertilizers in this way is creating lot of problems Viz; soil depletion of macronutrients and micronutrients, soil pollution, water and air pollutions."

III. Gaseous Air Pollution

It is somewhat disconcerting to the inorganic chemist to learn that with the exception of incompletely burned hydrocarbons, the common pollutants of the atmosphere are inorganic molecules. Although there are other sources, they commonly result from combustion processes.

Roasting processes in the preparation of metal sulfide ores can be a serious source of pollution if no effort is made to trap the sulfur dioxide by-product:

$$Cu_2S + 2O_2 \rightarrow 2\,CuO + SO_2 \quad (11.3)$$

One of the most serious problems in urban areas is the formation of smog and related forms of pollution from nitrogen oxides. The latter form in combustion processes, chiefly those occurring in internal combustion engines. A simplified scheme for the inorganic photochemistry of smog is:

$$N_2 + O_2 \rightarrow 2NO \text{ (engine cylinder)} \quad (11.4)$$

$$2NO + O_2 \rightarrow 2NO_2 \quad (11.5)$$

$$NO_2 + h\nu \rightarrow NO + O \qquad (11.6)$$

$$O + O_2 \rightarrow O_3 \qquad (11.7)$$

The ozone produced in this reaction is a strong oxidant and irritant. The above set of reactions can be further complicated by the formation of free radicals from incompletely burned hydrocarbons.

A product of civilization that is not normally considered to be a pollutant is carbon dioxide. The per capita consumption of energy is a good index of the extent to which civilization has advanced technologically, and most methods of generating power, such as steam-driven electrical generating plants and internal combustion engines, produce carbon dioxide from burning fossil fuels. Since carbon dioxide is a natural component of the atmosphere, direct deleterious effects to life are not expected (plants grow better in increased carbon dioxide concentrations). Instead natural processes should take place to "buffer" the concentration of carbon dioxide in the atmosphere. For example, for a given amount of carbon dioxide added to the atmosphere, about five-sixths should dissolve in the ocean. There it should slowly react with limestone sediments:

$$CO_2 + H_2O + CaCO_3 \rightarrow Ca^{2+} + 2HCO_3^- \qquad (11.8)$$

to "fix" the carbon dioxide. Further consumption of carbon dioxide results from weathering of exposed rocks on land. All of these processes are slow, however, compared to the production of carbon dioxide from combustion.

IV. Greenhouse Effect and Global Warming

The *greenhouse effect* was first discovered by Joseph Fourier in 1824, and first investigated quantitatively by Svante Arrhenius in 1896. *It is the process in which the absorption of infrared radiation by an atmosphere warms a planet.*

The name comes from an incorrect analogy with the way in which greenhouses used for gardening are heated by the sun in order to facilitate plant growth. In addition to the Earth, Mars, Venus, and other celestial bodies with atmospheres (such as Titan) have greenhouse effects. There are two types of greenhouse effects:

Natural greenhouse effect due to naturally occurring greenhouse gases.

Anthropogenic greenhouse effect resulting from gases emitted as a result of human activities.

Mechanism of Greenhouse Effect

The basic mechanism behind greenhouse effect is the fact that the earth's temperature is determined by the requirement that it produces the infrared flux needed to balance the absorbed solar flux. The Earth receives energy from the Sun in the form of radiation. To the extent that the Earth is in a steady state, the energy stored in the atmosphere and ocean does not change in time, so energy equal to the incident solar radiation must be radiated back to space. Radiation leaving the Earth takes two forms: reflected solar radiation and emitted thermal infrared radiation. The Earth reflects about 13% of the incident solar flux; the remaining 70% is absorbed, warms the land, atmosphere and oceans, and powers life on this planet.

The key to the greenhouse effect is the fact that the atmosphere is relatively transparent to visible solar radiation but strongly absorbing at the wavelengths of the thermal infrared radiation emitted by the surface and the atmosphere. The temperature of the atmosphere generally decreases with height above the surface, at a rate of roughly 6.5°C per kilometer on average, until one reaches the stratosphere 10–15 km above the surface. The more opaque the atmosphere, and the higher the emission level of the escaping infrared radiation, the warmer the surface.

Greenhouse Gases (GHG's)

Quantum mechanics provides the basis for computing the interactions between molecules and radiation. Most of this interaction occurs when the frequency of the radiation matches that of the spectral lines of the molecule, determined by the quantization of the modes of vibration and rotation of the molecule. The molecules / atoms that constitute the bulk of the atmosphere; oxygen (O_2), nitrogen (N_2), and argon (Ar) do not interact with infrared radiation significantly. While the oxygen and nitrogen molecules can vibrate, because of their symmetry, these vibrations do not create any transient charge separation that enhances the interaction with radiation. In the Earth's atmosphere, the dominant infrared absorbing gases are water vapor, carbon dioxide, and ozone, these molecules being "floppier" so that their rotation /vibration modes are more easily excited.

For example, carbon dioxide is a linear molecule, but it has an important vibrational mode in which the molecule bends with the carbon in the middle moving one way and the oxygens on the ends moving the other way, creating some charge separation, a dipole moment. A substantial part of the greenhouse effect due to carbon dioxide exists because this vibration is easily excited by infrared radiation. Clouds are also very important infrared absorbers. Therefore, water has multiple effects on infrared radiation, through its vapor phase and through its condensed phases. Other absorbers of significance include methane, nitrous oxide, and the chlorofluorocarbons.

Anthropogenic Greenhouse Effect

Carbon dioxide production from increased industrial activity (fossil fuel burning) and Measurements of CO_2 amounts from other human activities such as cement production and tropical deforestation has increased the CO_2 concentration in the atmosphere. Measurements of CO_2 amounts from Mauna Loa observatory shows that CO_2 has increased from about 313 ppm (parts per million) in 1960 to about 375 ppm in 2005. Because it is a greenhouse gas, elevated levels increase global mean temperature. There has been an observed global average temperature increase of about $0.5°C$ since 1960. CO_2 is transparent in the visible and near ultra violet regions of the spectrum in which the radiation reaching the earth from the sun lies.

On the other hand, the wavelength of radiation emitted by earth (acting as a black body) is considerably longer, and some of this infrared radiation is absorbed by the CO_2 in the atmosphere, holding the energy in the atmosphere rather than allowing it to be dissipated into space. Since the temperature of the earth is a balanced equilibrium between the energy received from the sun and that radiated back into space, such absorption by CO_2 could affect the earth's climates, resulting in the melting of the polar icecaps and concomitant problems. Although there is considerable debate about the *effect* and the *rate* of CO_2 accumulation, there is general agreement that the phenomenon is real.

V. Acid Rain

The combustion of high-sulfur fuels results in the emission of sulfur oxides into the atmosphere. In addition, some nitrogen in the air combines with oxygen upon passing through a flame, the more so the hotter the flame. These oxides eventually form sulfuric and nitric acids. Most industrialized nations (and those downwind of them) now experience a rainfall of low pH. The usual criterion for "acid rain" is having a pH below 5.6. The Los Angels basin routinely has fogs with water droplets of pH 2.2–4.0. As a consequence of the acid precipitation, marble statuary and buildings, often of great artistic and historical value, are being corroded and etched at an alarming rate.

Acid rain may have harmful effects on the ecosystem as well. If it falls in a limestone area, the buffering action of the calcium carbonate tends to neutralize its adverse effects. Most other rocks and soils lack this buffering action, and the pH of the water in lakes may drop below 5. Not only is the low pH harmful *per se*, but the acidified water dissolves metal oxides in the soil and releases zinc, manganese, and aluminium, all toxic if present in too great a quantity. Unlike naturally occurring sphagnum bogs / lakes, which have unusual but thriving flora and fauna, the artificially acidified lakes lack the high concentration of organic matter that provides humic acid to chelate these metals. (The presence of organic chelates of iron in surface waters has been related to the "red tide". The latter is an explosive "bloom" of algae (*Gymnodium breve*) that results in mass mortality of fish.)

VI. Ozone Depletion

One of the most important and controversial sets of atmospheric reactions at present is that revolving around stratospheric ozone. The importance of ozone and the effect of ultraviolet (UV) radiation on life is tremendous. A small portion of the sun's spectrum reaches the surface of the earth and that parts of the UV portion that are largely screened can cause various ill effects to living systems.

The earth is screened from far-UV (extremely high energy) radiation by oxygen in the atmosphere. The UV radiation cleaves the oxygen molecule to form two free radicals (oxygen atoms):

$$O_2 + h\nu \text{ (below 242 nm)} \rightarrow \cdot O \cdot + \cdot O \cdot \quad (11.9)$$

The oxygen atoms can then attack oxygen molecules to form ozone:

$$\cdot O \cdot + O_2 + M \rightarrow O_3 + M \quad (11.10)$$

The neutral body M carries off some of the kinetic energy of the oxygen atoms. This reduces the energy of the system and allows the bond to form to make ozone. The net reaction is therefore:

$$3O_2 + h\nu \rightarrow 2O_3 \quad (11.11)$$

This process protects the earth from the very energetic, short wave-length UV radiation and at the same time produces ozone, which absorbs somewhat longer wave-length radiation (moderately high energy) by a similar process:

$$O_3 + h\nu \text{ (220–320 nm)} \rightarrow O_2 + \cdot O \cdot \quad (11.12)$$

The products of this reaction can recombine as in Eq. 12.10, in which case the ozone has been regenerated and the energy of the UV radiation has been degraded to thermal energy. Alternatively, the oxygen atoms can recombine to form oxygen molecules by the reverse of Eq. 12.9, thereby reducing the concentration of ozone. An equilibrium is set up between this destruction of ozone and its generation via Eq. 12.11 and so under normal conditions the concentration of ozone remains constant.

The controversy over supersonic transports (SSTs) of the Concorde type revolves around the production of nitrogen oxides whenever air containing oxygen and passes through the very high temperatures of a jet engine.

One of these products, nitric oxide, reacts directly with ozone, thereby reducing its concentration in the stratosphere:

$$NO + O_3 \rightarrow NO_2 + O_2 \quad (11.13)$$

Furthermore, nitrogen dioxide formed in Eq. 12.13 or directly in the combustion process can react to scavenge oxygen free radicals and prevent their possible recombination with molecular oxygen to regenerate ozone (Eq. 12.10):

$$NO_2 + \cdot O \cdot \rightarrow NO + O_2 \quad (11.14)$$

Note that a combination of reactions 12.13 and 12.14 results in the net conversion of ozone to oxygen:

$$\cdot O \cdot + O_3 \rightarrow 2O_2 \quad (11.15)$$

And that the nitrogen oxides, either NO or NO_2, continuously recycle and thus act as catalysts for the decomposition of ozone:

$$\begin{array}{ccc} O_3 & NO & O_2 \\ \downarrow & \downarrow & \uparrow\uparrow \\ O_2 & NO_2 & \cdot O \end{array} \quad (11.16)$$

The current controversy revolves around the extent to which nitrogen oxides, NO_x, would be formed by SSTs and how much the ozone concentration would be affected.

The Ozone question is complicated by the fact that other chemicals are implicated in its destruction. Chlorofluorocarbons are widely used as propellants in spray cans and as refrigerants. They are extremely stable and long-lived in the environment. However, they too can undergo photolysis in the upper atmosphere:

$$F_3CCl + h\nu \ (190\text{–}220 \text{ nm}) \rightarrow F_3C\cdot + Cl\cdot \qquad (11.17)$$

The chlorine free radical can then interact with ozone in a manner analogous to the NO_x process:

$$Cl + O_3 \rightarrow ClO + O_2 \qquad (11.18)$$

$$ClO + O \rightarrow Cl + O_2 \qquad (11.19)$$

for a net reaction of:

$$O + O_3 \rightarrow 2O_2 \qquad (11.20)$$

with regeneration of the atomic chlorine. The chlorine thus acts as a catalyst and present evidence indicates that the ClO_x cycle may be three times more efficient in the destruction of ozone than the NO_x cycle.

VII. Other Industrial Pollution

There are many less visible sources of pollution. Dumping of waste acid baths, used electrolytic baths (containing cyanide), and various pollutants all contribute to poor water quality. Even more insidious is the release of relatively small amounts of toxic materials which may be concentrated in living organisms. The recent discovery of mercury in fish intended for human consumption is an example. It had been known that mercury was being lost from electrolytic cells in the manufacture of sodium hydroxide ("caustic soda") and chlorine, but the amount lost is small (of the order of 0.1 ppm in the waste water, which is further diluted upon discharge into a river).

It has now been found that microorganisms are able to transform it into methyl mercury cation. The latter is strongly held in the biological system (as might have been supposed from the binding of mercury by sulfur in proteins) and concentrated in the food chain. In view of the known effects of mercury on enzyme systems, this provides a particularly serious contamination of the ecosystem. It is merely another example of the principle that dilution does not solve pollution problems since living organisms can concentrate pollutants again to dangerously high levels.

VIII. Mining Problems

The process of winning a metal from its ores and using it in a technological application is · potential source of pollution even aside from the various purification, smelting, and manufacturing processes. Life has evolved and adapted to the availability of elements in the earth's crust. Basically every time man exploits relatively rare deposits of concentrated metal compounds a potential problem exists. Mining metals such as Co, Cu, Zn, Pb, and Hg and making no effort to prevent these elements from entering the ecosystem providing the possibility of a drastic overload of the biological systems affected. In some cases, such as iron, no problem is posed fro extravagant use of the metal. In other cases the amounts that could potentially enter the ecosystem are far larger than the "natural amounts" that are available from weathering of rock, recycling of organic matter, etc.

Although estimates of both the "natural" and man-made amounts of metals involved are difficult to make, the best estimates indicate, for example, that for the elements Ag, Au, Cd, Cr, Cu, Hg, Pb, Sb, Tl, and Zn the amounts entering the ocean yearly are from 4 to 600 times the amount lost from the ocean, and thus these elements present particularly likely pollution problems.

A second problem associated with the exploitation of mineral resources is that even after the mining activities cease, problems may still exist. For example, coal mines, both active and abandoned, release the equivalent of 8 million metric tons of sulfuric acid annually which pollutes over 15000 km of streams in Appalachia alone. The sulfuric acid results from the oxidation of iron sulfide both chemically and biochemically by chemolithotropic bacteria.

$$FeS \text{ (or } FeS_2) \rightarrow \tfrac{1}{2} Fe_2 O_3 + (2H_2 SO_4) H_2O \qquad (11.21)$$

Conclusion

It is highly unlikely that man (or any other organism) will, after adapting for millions of years to the concentrations of elements available in the earth's crust, suddenly benefit from random addition of various elements to the ecosystem.

A man must be conscious of his ecosystem. He must appreciate the following lines:

"I say that it touches a man that his blood is sea water, that the seed of his loins is scarcely different from the same cells in a sea-weed, and that of stuff like his bone is coral made. I say that physical and biological law lies

down with him, and wakes when a child stirs in the womb, and that the sap in a tree, up-rushing in the spring, and the smell of loam, where the bacteria bestir themselves in darkness, and the path of the sun in the heaven, these are facts of first importance to his mental conclusions, and that a man who goes in no consciousness of them is a drifter and a dreamer, without a home or any contact with reality."

–Donald Culross Peattie

PART II
HISTORY OF ATOM

"The fastest and most reliable way to master any science is to follow through its whole Path of development yourself."

–Felix Klein

"IMAGINATION
Will often carry us
TO WORLDS
That never were,
But without it
WE GO
NOWHERE"
–CARL SAGON

CHAPTER 1

1. Introduction

Atomic energy, radioactive isotopes, semi-conductors, elementary particles, masers, lasers: all quite familiar terms, yet the oldest is hardly forty years of age. They are all children of 20th century, the century of science. Knowledge is advancing at a fantastic rate, and every new step opens up new vistas. The old sciences are going through a second youth.

An enormous tree of knowledge has grown out of the seminal ideas expressed by Planck, which served as a starting point for amazing discoveries. Out of Planck's concepts grew quantum mechanics, which opened up an entirely new world—the world of the ultra-small, of atoms, of atomic nuclei and elementary particles.

In a way, people did guess and conjecture about this atom before 20th century. The inquisitive human mind had speculated upon these things and had long imagined reality only many centuries later. In ancient times, long before the first travelers laid their paths of discovery, man had guessed that there were people and animals and land beyond the little area in which he lived.

In the same way, people felt that there existed a world of the ultra-small long before it was actually discovered. One did not need to go far in search of this new world, for it was right at hand, lying around him in all things. In olden times, thinkers had meditated on the way nature had produced the world around us out of something quite formless.

The Creation Hymn 10.129 of Great Hindu Scripture "The RgVeda" mentions in verse 1:

CHEMISTRY AND OUR DAILY EXISTENCE

"There was neither non-existence nor existence then; there was neither the realm of space nor the sky which is beyond. What stirred? Where? In Whose Protection? Was there water, bottomless deep?"

How was it, they queried, that it came to be inhabited by its great diversity of things. Might it not be that nature worked like a builder that makes large houses out of small stones? Then what are these stones? Enormous mountains are weathered away by the water, the wind and mysterious forces. The rocks that came away are in time broken down into small pieces. Hundreds and thousands of years pass, and these are pulverized into dust.

Is there a limit to this dividing and sub-dividing of matter? Are there particles so small that even nature is no longer able to break them up? The answer was 'Yes'. This is what the ancient philosophers Epicurus, Democritus and others said. These particles are given the name "ATOM." Their chief property was that no further division is possible. The word 'atom' in Greek means 'indivisible'.

The conception of atoms in an age when science was still in its infancy was a conjecture of genius. But this conjecture did not follow from any kind of observations and was not supported by any kind of experiments. The atoms were forgotten for a very long time.

Atoms were recalled rather invented once again only at the beginning of the 19th century by chemists. The Frenchman Lavoisier and the Englishman Dalton demonstrated that chemistry is capable of penetrating deep into the essence of things. The chemists, physicists, and mathematicians of that time made a whole series of discoveries that prepared the way for the flourishing of the exact sciences in the latter half of the 19th century.

Many great minds during the Renaissance were engaged in problems that later formed the basis of classical mechanics. However, as the edifice of classical physics grew upwards, its enormous front gave signs of fatigue, sinister cracks, and finally the entire structure began to crumble under the bombardment of new facts. Out of these cracks in the structure of classical mechanics grew the theory of relativity and quantum theory.

Quantum Mechanics was born at the turn of the century. In view of the dual nature of matter, quantum mechanics was refined to 'Wave mechanics'. But here again we have only half of it—there is no mention of quanta. The introduction of new terms in science is a laborious and thankless task. New terms come in slowly and change still more slowly. Physicists understand the new meaning that these terms carry and it is for us to learn them.

2. Democritus' Apple

The idea of the atom is usually considered to have been first proposed by the ancient Greek philosopher Democritus, though history also mentions his teacher Leucippus of Miletus (or perhaps Elea) and the ancient Hindu philosopher Kanada who lived just before the beginning of the Christian era and was one of the early atomists.

Kanada also known as Kashyapa was an ancient Indian natural scientist and philosopher who founded the *Vaisheshika* school of Indian philosophy. Estimated to have lived sometime between 6th century to 2nd century BCE, "Kanada", his traditional name means "feeder on atoms." He is known for developing the foundations of an atomistic natural Indian philosophy in the Sanskrit text *Vaisheshika Sutra*. His text is also known as Kanada Sutras, or aphorisms of Kanada.

The school founded by Kanada attempted to explain the creation and existence of the universe by proposing an Atomistic theory, applying logic and realism, and is among one of the earliest known systematic realist ontology in human history. Kanada suggested that everything can be subdivided but this subdivision cannot go on forever, and there must be smallest entities (*parmanu*) that aggregate in different ways to yield complex substances and bodies with unique identity, a process that involves heat, and this is the basis for all material existence.

"According to Kanada's Sutras (IV, I), an atom is 'something existing, having no cause, eternal' (*sad akaranavan nityam*). They are moreover, described as less than the least, invisible, intangible, indivisible, imperceptible, by the senses; and what is most noteworthy in distinguishing the *Vaiseshika* system of philosophy from others—as having each of them a *Vaisesha* or eternal essence of its own.......... One might even be tempted to contrast some of the discoveries of modern chemists and physicists with the crude but shrewd ideas of Indian philosophers prosecuting their investigations more than 2000 years ago without the aids and applications now at everyone's command."

According to Kanada, the infinite divisibility of matter is absurd because ".....the infinite is always equal to the infinite." The tiniest particle in nature is a speck of dust in a sunbeam. It consists of six atoms linked pairwise "by the Will of God or for some other good reason."

Our knowledge about Democritus is very scarce. Democritus was an ancient Greek pre-Socratic philosopher primarily remembered today for his formulation of an atomic theory of the universe. Democritus was born in

Abdera, Thrace, around 460 BC, although there are disagreements about the exact year.

We know from the legend that Democritus once sat on a stone by the seashore, held an apple in his hand and reasoned; "If I cut this apple in half I will have two halves, if I cut one half in two I will have two fourths. Now if I continue to divide the remaining parts in the same way, will I continue to obtain the eighth, a sixteenth, a thirty second, etc. of an apple? It turned out subsequently that Democritus' doubt (as do almost all unselfish doubts) contained a grain of truth. On second thought, the philosopher came to the conclusion that a limit to such divisibility exists. He named the last, already divisible particle an 'atom' (from the Greek 'Atomos' meaning indivisible). He set down his views in a book called the 'Little World System'. The following are fragments what he wrote more than two thousand years ago:

> "The universe is made up, in reality of nothing but atoms and the void; all the rest exists only in the mind. There are countless worlds and each has a beginning and an end in time. Atoms are innumerable in size and quantity, moving in all directions in a void. They are indestructible and unchangeable owing to their hardness."

3. After Long Oblivion

Aristotle did not believe in the atomic theory developed by Democritus and he taught so otherwise. He thought that all materials on Earth were not made of atoms, but of the four elements: Earth, Fire, Water and Air. He believed that all substances were made of small amounts of these four elements of matter. Aristotle was opposed to the idea of atomism.

Democritus was forgotten for many centuries and his works were painstakingly destroyed. This is why his teachings have come down to us only in fragments and attestations of his contemporaries. They were not accepted, however by other philosophers and primarily by his contemporary Aristotle, the future teacher of Alexander the Great. When Democritus died, Aristotle was only fourteen years old. Democritus and his works became known in Europe from the poem called "*De Rerum Natura*" ('on the nature of things') by Titus Lucretius Carus (c. 95–55BC).

Physics as a science came into existence in the 17th century and it soon supplanted ancient natural philosophy. The new science was based on the experiments and mathematics rather than on pure speculation. People began to study Nature around them instead of merely observing it. They began to conduct deliberate experiments to check various hypotheses and to record the results of these tests in the form of numbers. Aristotle's idea could not pass such a test. Democritus' hypothesis could, though almost nothing has remained of its initial substance.

After 20 centuries of oblivion, the idea of atoms was restored to life by the French philosopher Pierre Gassendi (1592–1655) for which he was persecuted by the church. During the Middle Ages, scientists were not only persecuted for various hypotheses, but also for rigorous facts if these facts contradicted universally recognized dogmata. Nevertheless, the atomic hypothesis was accepted by all the most advanced scientists of the time.

Even Newton, with his famous motto *"Hypotheses Non Fingo"* ('I frame no hyothesis') believed in it in his own way at the end of his *Optiks*. However, notwithstanding its attraction, the hypothesis unverified by experiments was doomed to remain only a hypothesis.

The first clear proof that Democritus was right, rather than Aristotle, was found by the Scottish botanist Robert Brown (1773–1858). In the summer of 1827, Brown observed that the finest pollen grains of the plant *Clarkia pulchella* have an irregular motion when suspended in water due to the actions of some unknown force. He soon wrote and published a paper on this matter. At first his experiment gave rise to perplexity, which was worsened by Brown himself when he tried to explain the phenomenon as the result of some "Vital Force" inherent in all organic molecules. Naturally, such an artless explanation of the Brownian motion could not satisfy scientists, and they continued their investigations. Especially persistent were the Belgian Father Ignace Carbonelle (1880) and the Frenchman Louis Georges Gouy (1888). Nineteen centuries before Brown, these properties of the Brownian motion were first pictured by the imagination of Lucretius and described in detail in his famous poem.

This strange motion did not initially attract the attention it deserved. Most physicists had heard nothing about it and those who had, considered it to be devoid of interest because they held it to be similar to the motion of specks of dust in sunbeam. It took about 40 years, probably for the idea to take shape that the random motions of plant pollen seen under a microscope were due to chance impacts of tiny invisible particles of the liquid in which the pollen grains were suspended. Almost everybody was convinced of this when Gouy's work was published, and the atomic hypothesis acquired numerous supporters.

Many people were quite sure, even before brown's time that all bodies are buildup of atoms. Certain properties of atoms were obvious to them without any further investigations. Notwithstanding the enormous differences found among them, all bodies in nature have weight and size. Evidently their atoms must also have weight and size. Precisely these properties of atoms formed the basis for the reasoning of John Dalton.

CHAPTER 2

1. Scientific Basis of the Constitution of Matter

The scientific basis of the constitution of matter was established by John Dalton (1766–1844). He was an unassuming teacher of mathematics and natural philosophy at New College in Manchester and a great scientist who determined the development of Chemistry for approximately the next hundred years.

One question that immediately arose before converts to atomism was: Does the great diversity of bodies signify as great a diversity of atoms as maintained by Democritus? This proved to be wrong.

John Dalton after investigating chemical reactions in detail, first clearly formulated in 1808 the definition of a chemical element: <u>An element is a substance that consists of atoms of a single kind</u>.

He put forward his famous Atomic Theory and based it on the firm foundation of the laws of Chemical Combination. Among the scientists of his time, John Dalton was a very unique figure. By the beginning of 19th century everyone had come to believe in science and understood the secret of its power: it dealt with numbers and numbers would never deceive you. That is why the art of conducting precise experiments was valued higher than any other capability in those times.

Dalton absolutely lacked this skill, and consequently was subject to severe attacks from his more dignified colleagues and all the vulnerable scientists of his day. With respect to his cast of mind, Dalton was a typical theoretician, as we now picture this profession. Hence the inaccuracy in his works should not be judged too strictly. On the basis of these works he expressed lucid and fruitful ideas. The gist of his discovery is that he indicated an experimental way to verify the atomic hypothesis.

According to Dalton the atoms of different substances differ in weight and remain unchanged in all transformations of the substances; they are only re-arranged. The modern history of the atom begins with Dalton. He was the first who not only firmly believed in the atomic hypothesis, but began to search for its consequences, especially observable ones. His line of reasoning was approximately the following:

Assume that all elements consist of atoms. Then say, 16 grams of oxygen contain N atoms of oxygen. Now assume, further that we burn Hydrogen in this oxygen. It can easily be measured that we need 2 grams of hydrogen to burn 16 grams of oxygen and as a result we obtain 18 grams of water.

The first supposition that would occur to any supporter of the atomic hypothesis is that each atom of oxygen O combines with one atom of hydrogen H and forms one molecule of water HO. This is exactly what Dalton thought.

Subsequently Berzelius proved that he was wrong about water and that each atom of oxygen combines with <u>two</u> atoms of hydrogen so that the chemical formula of water acquires its familiar form: H_2O. But the important point here is the idea that a whole number of atoms of hydrogen combines with each atom of oxygen. Hence, if 16 grams of oxygen contain N atoms, then 2 grams of hydrogen contain 2N atoms. This means: <u>One atom of oxygen is 16 times heavier than an atom of hydrogen</u>.

Thus it became possible to compare the weights of atoms of different elements. This led to a new concept, the <u>atomic weight</u>, which is simply a number indicating how many times the weight of an atom of some element is heavier than an atom of hydrogen, whose atomic weight is arbitrarily equal to 1. Thus numbers were first made use of in the science of the atom. This was an event of extraordinary importance.

In 1808 Dalton published his famous book <u>A New System of Chemical Philosophy</u>. It ushered in a whole new age in science. His conclusions were immediately verified by the English physician and chemist William Hyde Wollaston (who had first discovered the dark lines in the solar spectrum) and he found them to be quite correct.

It is difficult for us today to picture that confused age when not only the atomic hypothesis was repudiated, but it was even doubted that chemical compounds had a constant composition. One of the events of this period was famed eight-year controversy between Joseph Louis Proust and Claude Louis Berthollet which ended only after Proust had finally proved that ir-regard-less of how a compound is obtained, it always has the same fixed composition.

It remained to take the last step: to learn to determine the atomic weights of elements. For this purpose, it was necessary to start with the simplest substances. First scientists turned their attention to the gases, very soon in 1809, the French chemist Joseph Louis Gay-Lussac (1778–1850), known as the author of Gay-Lussac's Law, found that the volumes of two gases that combine in a chemical reaction are always in small whole number ratios. Not the weights, as we see, but the <u>volumes</u>.

This is extremely important as it leads to the conclusion that equal volumes of gases contain equal number of atoms. This was precisely the conclusion reached in 1811 by the Italian scientist Amedeo Avogadro (1776–1856) except that his wording was more exact: <u>Equal volumes of gases contain equal number of molecules</u>.

The number N of molecules contained in 22.4 liters of any gas, we now called <u>Avogadro's Number</u>. This is one of the main universal physical constants like the velocity of light c and Planck's constant h. The number N could first be evaluated after the calculations made by Joseph Loschmidt (1821–1895).

The work of Joseph Loschmidt, a physics teacher of the Vienna University should be considered the first really successful endeavor to determine the size and mass of atoms. In 1865, he found that the size of all atoms is ~ the same and is equal to 10^{-8} cm and that an atom of hydrogen weighs only 10^{-24} gms.

Avogadro's hypothesis was soon forgotten, and only half a century later, in 1858, it was revived by another Italian scientist, Cannizzaro (1826–1910). This happened most opportunely because the chemists of that time could not come to any general agreement. Each one recognized only his own table of atomic weights. The organic chemists did not trust the inorganic chemists, and the first International Chemical Congress, a meeting of the world's most famous chemists held in Karlsruhe, Germany, in 1860, could not reach any satisfactory agreement on these matters. Although, in a resolution dated September 4, 1860, the Congress did establish the difference between atoms and molecules.

2. Doctrine of Elements

Philosophers of the Ionian School, Whose famous representative was Thales of Miletus (c. 640–546 B. C.) recognized only one basic element—water—"on which the earth floats and which is the beginning of all things". Later Empedocles of Agrigentum (c. 490-c. 430 B.C.) added three more elements—earth, fire and air—to water. Finally Aristotle (384–322 B.C.) added a fifth essence of which the heavenly bodies were supposed to have been made, to the other four elements.

Vaiseshika system of Hindu philosophy, attributed to an author Kanada is certainly the most interesting of all the systems, both from its practical character and from the parallels it offers to European philosophical ideas. It begins by arranging its inquiries under seven *Padarthas,* which as they are more properly categories. They are as follows: 1. Substance (*dravya*). 2. Quality or property (*guna*). 3. Act or action (*karman*). 4. Generality or community of properties (*samanya*). 5. Particularity or individuality (*visesha*). 6. Co-inherence or perpetual intimate relation (*samaiaya*). 7. Non-existence or negation of existence (*abhava*).

Kanada, however, the author of the *Sutras*, enumerated only six categories. The seventh was added by later workers. The first category of *Dravya* or 'substace' is sub-divided into nine Dravyas, as mentioned in his fifth Sutra. These are: 1.Eearth (*prithivi*), 2. Water (*apas*), 3. Light (*tejas*), 4. Air (*vayu*), 5. Ether (*akasa*), 6. Time (*kala*), 7. Space (*dis*), 8. Soul (*atman*), 9. The internal organ, Mind (*manas*).

In the Middle Ages, the doctrine of elements was revived by the Alchemists, the most famous of which were the Egyptian Zosimus (~300A.D.), who wrote a 28-volume encyclopedia on chemical knowledge up to his day, The Arab Geber (760–815 A.D.), and Saint Albertus Magnus (c. 1193–1280).

The Alchemists conceived of the elements as being qualities or 'principles' and not substances. Mercury served as the principle of metallic luster, Sulphur—combustibility and salt—solubility. They were sure that if these 'principles' were mixed together in the proper proportions, any substance in nature could be obtained.

As a rule, the word "Alchemy" is associated with tales of attempts to change mercury into gold, the concocting of the elixir of life and other miracles. A little rummaging in the archives may turn up, for instance, a work of Geber in which he seriously discussed the question: "Why, as everyone knows, a cloud provides no rain when a naked woman comes out of her house and faces this cloud?"

But, in addition to this obvious non-sense, alchemists did invent alcohol which in itself is sufficient to justify their existence. Their main service to mankind, however, is that by their blind experiments, the alchemists were wont to practice, although with no adequate direction or knowledge, gradually led to the accumulation of facts without which the science of chemistry would never have been founded.

In the 17th century, alchemy and natural philosophy made way for chemistry and physics. In 1642, Joachim Jungius (1587–1657) published his <u>De Principis Naturalism</u>. He closes these <u>Disputes on the Principles of Matter</u> quite in the spirit of the century: "Which principles should be recognized as primary for homogeneous bodies is a question that can be answered only by honest, detailed and diligent observations rather than by making guesses."

A famous book, <u>The Sceptical Chymist</u> written by Robert Boyle, was published in 1661. In it he defines chemical elements as simple or primitive substances that cannot be broken down into or be produced by uniting simpler substances.

In essence, this is the first and almost modern definition of an element: An element is primarily a <u>substance</u>, and by no means a "principle", substratum or idea. What still remained vague was how to extract elements from natural substances, and how to distinguish pure elements from their mixtures or compounds. For instance, Boyle himself supposed that water was almost the only pure element, and at the same time, considered gold, copper, mercury and Sulphur to be chemical compounds and mixtures.

Antoine Laurent Lavoisier (1743–1794), the Father of Chemistry, completely accepted Boyle's ideas, but since he lived a century later, this was insufficient for him. He wanted to learn to separate elements out of chemical compounds. Evidently, he was one of the first to employ scales for purposes of investigation and not to prepare powders and mixtures. He proceeded from an assumption which seems trivial today, but which required no little courage in the age of phlosgiston: <u>Each element of a compound weighs less than the compound as a whole</u>.

Consequently, applying this principle, he drew up the first table containing about 30 elements. The views of Lavoiaier were so contradictory to the generally accepted ones that the zealous followers of the phlogiston theory in Germany burnt his portrait in public.

Lavoisier did not finish his investigations. Accused of treason, he was guillotined on the Place de la Revolution in the afternoon of May 8, 1794 and his body was buried in a common grave. Next morning, Joseph Louis Lagrange, the great mathematician said bitterly:

CHEMISTRY AND OUR DAILY EXISTENCE

"It required only a moment to sever that head, and perhaps a century will not be sufficient to produce another like it."

The next 100 years were marked by the works of chemists who gradually added new elements to Lavoisier's table. Especially worthy of admiration is the "king of Chemists", Baron Jons Jakob Berzelius (1779–1848) who analyzed over 200 substances and discovered several new elements. Incidentally it was he who introduced in 1814 the modern notation of the chemical elements, using the first letters of their Latin or Greek names.

By that time about 60 elements were known. This was not as many as Democritus had expected. It certainly seemed probable that the set of elements formed a unified <u>system</u> and the search began to find one.

In this connection many attempts were made to find a system of elements. Mention may be made of attempts made at various times by Marne in 1786, William Prout in 1815, Johann Wolfgang Dobereiner in 1817, Max Pettenkofer in 1850, John Gladstone in 1853, William Odling in 1857, Alexander Chancourtois in 1863, John Alexander Newlands in 1865 and many others.

At the basis of any science is the human capacity to wonder at things. The existence of elements always was and always will be a cause of wonderment. In the minds of scientists, however, the feeling of wonderment is soon followed by a pressing need to put the impressions they received into some kind of order. This is a pure human trait. It is deep within each one of us. A child is overjoyed when he manages to make a regular figure out of a chaos of building blocks, a sculptor when he carves a statue from a block of marble.

The first question that arises when we want to put things into some sort of order is: What is the underlying feature? Any classification makes sense only if it enables fundamental properties or structural features to be revealed.

Chemical elements have very many properties. This is quite understandable; otherwise they could not make up this diversified world of ours.

The many unsuccessful attempts to find a system of elements helped scientists to realize finally that among the various properties of elements that can be directly observed not a single one was suitable for the basis of their classification. Finally the matter of atomic weight was sufficiently clarified, and the atomic weights of the elements could be determined correctly enough to begin their classification. The required property, <u>the atomic weight</u> lies outside the dominion of chemistry; it belongs completely to physics. The moment when this was realized can be regarded as the starting point of the modern theory of chemical elements. This decisive step was taken by John Dalton.

Russian chemistry professor Dmitri Mendeleev published his Periodic table of elements in 1869, using atomic weight to organize the elements, information determinable to fair precision in his time. Atomic weight worked well enough to allow Mendeleev to accurately predict the properties of missing elements. Mendeleev constructed his periodic table just 5 years after John Newlands put forward his Law of Octaves. Mendeleev constructed his Periodic Table in the order of increasing atomic weights, into horizontal rows called Periods, and vertical columns called Groups.

CHAPTER 3

1. Birth of Spectral Analysis

The history of physics is not simply a collection of facts, but a coherent picture of the origin and development of physical ideas, without which science may seem to be a random set of formulas and concepts. The history of physics is a necessary element in the introduction of a physicist. Without it he will remain a mediocre scientist all his life. To understand the completeness and elegance of the concepts of modern physicist is necessary to retrace their sources and their ways of development.

If we follow a sunbeam observantly, it can lead us right up to the threshold of quantum physics. If we pass a ray of sunlight through a prism, a spectrum will appear on a screen behind the prism. This is a familiar phenomenon; we have become used to it in 200 years. On the face of it there are no sharp boundaries between the different parts of this spectrum: red gradually goes over to orange, orange to yellow etc. Credit for the discovery of spectra goes to the diversified genius of Newton.

That is what everyone thought until in 1802, the English physician and chemist, William Hyde Wollaston (1766–1828), examined this spectrum more intently. For this purpose he built the first spectrograph with a slit and with it he discovered several distinct dark lines which crossed the solar spectrum without any apparent order at various places. He attached no importance to these lines. Subsequently, they were named Fraunhofer Lines after their real investigator, and not their discoverer.

Joseph Von Fraunhofer (1787–1826) ground optical glass and painstakingly investigated the dark lines in the solar spectrum. He found 574 lines, labelled the most prominent ones by letters of the alphabet and indicated their exact location in the spectrum. Their positions were strictly invariable.

A sharp double line, in particular, called the D line by Fraunhofer, always appeared at the same place in the yellow region. Fraunhofer established another important fact: he found a <u>bright double yellow line</u> in the spectrum of the flame of a spirit lamp which always occupied exactly the same place as the dark D line in the solar spectrum.

The significance of this fact was appreciated only after many years had passed. Of Fraunhofer's discoveries, the most important to us just now are his observations of the double D line. Then, in 1814 when he published his investigations, no special attention was paid to his observations. His work was not in vain, however; 43 years passed and William Swan (1818–1894) established that the double yellow D line in the spectrum of a spirit lamp flame appears only in the presence of the metal sodium. (Its traces as a component of common salt can almost always be found in various substances, as well as in a spirit lamp).

As many other scientists before him, Swan did not realize the significance of his discovery and therefore did not say the decisive words: "This line belongs to the metal sodium." This simple and important idea came only two years later (in 1859) to two professors; Gustav Robert Kirchoff (1824–1887) and Robert Wilhelm Bunsen (1811–1899). Both of them correctly concluded that the role of the glass prism consisted only in sorting the incident rays of light into their wavelengths. The extended band of the solar spectrum indicated that all the wavelengths of visible light were present. The yellow line, which appeared when the light source was a burning rag, indicated that the spectrum of table salt had a single specific wavelength.

It was proved beyond doubt that the yellow line, obtained by using a Bunsen Burner belonged to sodium. Subsequently, this modest observation of the double D line of sodium led to the birth of <u>Spectral Analysis</u>.

Later, it was found that sodium is no exception in this respect. Every chemical element has its own characteristic spectrum. As a rule, some of the spectra were much more complicated than that of sodium and consisted at times of a very larger number of lines. But no matter what the compound or substance the element appeared in, its spectrum was always distinct, like the photograph of a person.

We now know two kinds of spectra: continuous (or thermal) and line spectra. A thermal spectrum contains all wavelengths. It is radiated when solid bodies are heated and it does not depend upon the nature of the bodies.

A line spectrum consists of a set of sharply defined lines; it appears upon heating gases and vapors and what is especially important, the set of lines is unique for each element. Moreover, the line spectra of elements do not depend upon the kind of chemical compound made up of these elements. Hence we must look for an explanation of their spectra in the properties of atoms.

The fact that elements uniquely and quite definitely determine the kind of line spectrum observed was soon recognized by everybody that the same spectrum characterizes a <u>single atom</u> was not realized at once, but only in 1874 thanks to the work of the famous English Astro physicist Sir Joseph Norman Lockyer (1836–1920). Incidentally, the same ideas were expressed earlier by Maxwell in 1860 and Boltzmann in 1866. And they did realize it they immediately reached the inevitable conclusion: Since a line spectrum is originated <u>inside</u> a separate atom, the atom must <u>have a structure</u>, i.e. it must have <u>component parts</u>.

By the 20th century, hundreds of papers had appeared dealing with the respectable science of spectroscopy. Spectral analysis was moving quite ahead at quite a space doing great service in chemistry, astronomy, metallurgy and other sciences.

2. Discovery of Electron

In 1865, when Joseph Loschmidt's works appeared not much was known about atoms. They were pictured as small solid spheres about 10^{-8} cm in size and weighing from 10^{-24} to 10^{-22} grams. Each such sphere could be ascribed an atomic weight, i.e. a number indicating how many times it is heavier than an atom of hydrogen. The concept of atoms as being solid spheres was adequate to explain numerous facts from chemistry, theory of heat, and the structure of matter. By 1870, however, the idea that atoms consist of still simpler particles had already taken shape, and physicists started to look for them. First of all they investigated the <u>electrical properties</u> of the atom.

As a rule all substances are electrically neutral. Under certain conditions, however, they begin to display electrical properties, for instance, when you rub glass with wool, amber with silk etc. These properties are especially distinctly manifested in electrolysis phenomena. In 1834, Michael Faraday (1791–1867) established the quantitative laws of electrolysis. He was the first to use the name 'ion' for the charged atoms. The charge carried by one ion equals

$$e = 4.802 \times 10^{-10} \text{ esu}$$

This value is very small and no charges smaller than this elementary charge have ever been found. In 1891, the Irish physicist George Johnstone

(1826–1911) had the good fortune to suggest the name for this minimum possible charge that it is known by today. He called it an electron.

In the beginning, the concept of a particle was not linked to this term. It merely served to designate that minimum amount of charge that can be carried by an ion of any atom. The latent idea, however, that an electron is a particle, always existed. In 1881, Helmholtz (1821–1894) first clearly formulated the idea of the molecular structure of electricity.

He said "If we accept the hypothesis that the elementary substances are composed of atoms, we cannot avoid concluding that electricity also, positive as well as negative, is divided into definite elementary portions which behave like atoms of electricity."

As such of course, this idea was nt new even then. Way back in 1749, the great American statesman, scientist and inventor, Benjamin Franklin supported something like this, but his guess was not based on any known facts and therefore did not lead to any new consequences. In 1871, the German physicist Weber (1804–1891) returned to Franklin's idea, but met no sympathy. By his time, so much was already known about electricity that no hypothesis could be accepted without definite proof; knowledge presupposes responsibility.it was necessary to obtain experimental proofs of the idea of electrons. Attempts were made to find them in the conduction phenomena of gases:

Imagine a glass tube with some gas (e.g. Neon) and sealed at both ends together with wires (usually of platinum). If we connect these two wires to different poles of a battery, one to the negative pole (cathode), and the other to the positive pole (anode), current will flow in the circuit, exactly in the same manner as when we use an electrolyte. Probably, it was just this analogy with electrolysis that incited Faraday in 1838 to make the prototype of such a tube (Faraday's "electric egg").

About the middle of last century, Julius Plucker (1801–1868) became enthusiastic about experimental physics. He established that the conductivity of the gas depends upon the concentration in the tube and increases if a part of the gas is exhausted from the tube. Each gas begins to glow with its own particular color, so that the composition of the gas in the tube can be determined from its color.

If the tube is further evacuated a dark space appears near the cathode ("Faraday dark space"). A pupil of Plucker's, Eugen Goldstein (1850–1930) called this radiation cathode rays in 1876. Johann Wilhelm Hittorf (1824–1914) found that these rays were deflected in a magnetic field. Finally, in 1879, Varley 1828–1883) showed that they are negatively charged.

The English physicist and chemist William Crookes devised still better evacuated tube. Now another, even darker region called the Crookes dark space was produced at the cathode. It also grew gradually until it filled the whole tube after which the anode started to glow with a faint greenish light. That day in 1878, when this happened, can be regarded as the birthday of the cathode-ray tube, the main component of television sets.

The indisputable experiments of Crookes discovered the following amazing properties of cathode rays;

1. It travels in straight line.
2. It causes bodies to glow and can even melt them.
3. It is deflected in electric and magnetic fields.
4. Its mean free path in air is 7 cm while that of atoms is only 0.002 cm.
5. Cathode rays consist of a stream of negatively charged particles of a size much smaller than atoms.

John Perrin finally proved in 1895 that cathode rays are negatively charged. During the next two years it was found that their velocity is about one-tenth of the velocity of light. These and all other properties of these rays did not depend upon the composition of the gas in the tube. This meant that cathode particles are the indispensable components of all atoms.

Finally, in 1897, J.J. Thomson managed to measure the electric charge e and the mass m of a single "atom of electricity".

$$e = \sim 5 \times 10^{-10} \text{ esu}$$
$$m = \sim 10^{-27} \text{ gms}$$

The charge turned out to be almost exactly equal to the charge of an ion of hydrogen measure in investigating electrolysis.

The mass turned out to be only about one-thousandth of the mass of an atom of hydrogen.

The history of the electron is exemplary in understanding the logical sequence of discoveries made in modern physics. The hypothesis of the electron originated from the observations of Faraday, Plucker and Crookes.

The fertility of this hypothesis was tested and proved by the experiments of J. J. Thomson and other physicists. In 1900 Paul Drude (1863–1906) suggested that this particle be called the ELECTRON.

Physicists, who believed in the reality of electrons from the very start, carefully measured its characteristics: charge e and mass m. The present values of e and m of an electron are given below:

$$e = 1.6 \times 10^{-19} \text{ coulombs}$$

$$m = 9.1 \times 10^{-31} \text{ kgs}$$

J. J. Thomson was awarded a Nobel Prize in 1906 in recognition of the great merits of his theoretical and experimental investigations on the conduction of electricity by gases.

The schematic diagrams of cathode ray apparatus and their splitting of rays in electric and magnetic fields are given in Fig. I, II, and III respectively on next page.

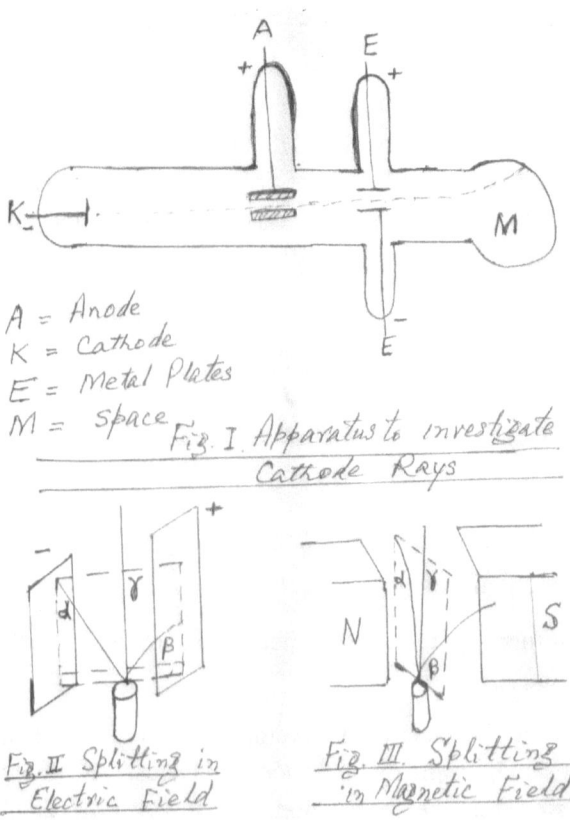

A = Anode
K = Cathode
E = Metal Plates
M = Space

Fig. I. Apparatus to investigate Cathode Rays

Fig. II Splitting in Electric Field

Fig. III. Splitting in Magnetic Field

Positive Rays

Goldstein in 1886 while carrying out experiments with a vacuum discharge tube provided with a perforated cathode observed luminous streamers issuing from the fine holes in the cathode into the region behind the latter i.e. on the side of the cathode opposite to the anode. The luminosity of the streamers is due to the phosphorescent effect they produce on the residual gas in the discharge tube.

Because of the observation that the rays issued from the fine holes or channels in the cathode, Goldstein called these rays Canal rays. Later J. J. Thomson gave them their present name—positive rays on account of the fact that they consist mainly of positive ions i.e. atoms and molecules of the gas that have lost one or more electrons.

The term "positive rays", however is not wholly correct, as the positively charged ions are accompanied by negatively charged ions and neutral molecules. In fact, there exists an equilibrium in which the ratio of the number of charged to uncharged particles under a given set of conditions, is constant.

Positive rays differ from cathode rays in many respects e.g. while cathode rays actually start from the cathode the positive rays appear to originate in the glow that surrounds the Crookes dark space.

Sir J. J. Thomson in 1911 devised a method of separating the charged particles into groups according to their respective e/m. The particles of a given group, though possessing different velocities, have the same e/m. This method of resolving a positive ray beam, called positive ray analysis has not only served its original purpose of determining the ratio e/m of the different particles constituting the positive ray beam but also led to the discovery of 'isotopes'.

CHAPTER 4

1. J. J. Thomson's Model

J. J. Thomson investigated the scattering of X-rays by the atoms of various elements. He came to the conclusion that there were comparatively few electrons in an atom and that their number approximately equals one half of the atomic weight of the element.

In 1904, he proposed his atomic model, which soon became known as "plum pudding atom", based on the hypothesis of his equally famous compatriot William Thomson (Lord Kelvin).

The 'plum pudding model' was put forth before the discovery of the nucleus. According to this model, the atom s a sphere of positive charge, and negatively charged electrons are embedded in it to balance the total positive charge. Thus on the whole, the atom is electrically neutral. J. J. Thomson himself showed no special enthusiasm for his model.

2. Planetary Model

George Johnstone Stoney surmised as far back as 1891 that the electrons move around within an atom like the satellites of planets. A Japanese Physicist Hantaro Nagaoka (1865–1950) contended in 1902 that the space inside an atom is extremely huge in comparison to the electric grains, or in other words, that the atom is in its way a complex astronomical system, similar to the rings of Saturn.

Lorentz and Larmor in 1896 had used the planetary structure of the atom to explain Zeeman's discovery of the splitting of spectral lines in a magnetic field.

The famous French physicist and mathematician Jules Henri Poincare (1854–1912) wrote quite as definitely that:

"All the experiments on the conductivity of gases....provide us with grounds for regarding the atom as consisting of a positively charged center, of a mass approximately equal to that of the atom itself, about which electrons attracted to this nucleus revolve".

3. Rutherford's Model

The Rutherford model is a model of the atom devised by Ernest Rutherford, a New Zealand born physicist. Rutherford directed the famous Geiger-Marsden experiment in 1909 which suggested upon Rutherford's 1911 analysis that J. J. Thomson plum pudding model of the atom was incorrect.

Their apparatus consisted of a vial containing radium-C, which emits alpha particles, (An alpha particle has a mass of 4 atomic mass units (mass of an electron is only 1/1820 of this unit), and a positive charge twice that of an electron in absolute value. In short, alpha particles are helium nuclei: $_2He^4$), a diaphragm which confines these particles to a narrow beam and directs them on to a Zinc Sulphide screen and a microscope for observing the scintillations of the alpha particles on the screen.

Now when a piece of metal foil is placed in the path of the alpha particles, a blurred band is obtained on the screen instead of the clear-cut image of the slit. It was found that the majority of alpha particles passed through the foil without deviation, some were deflected through various angles, and a certain very small fraction rebounded and almost retraced their original path.

The results of these experiments, especially the rebounds, could not be explained on the basis of Thomson's model. Indeed an alpha particle, which possesses a double positive charge, travels at a high velocity and has a relatively large mass, can be sharply thrown back only if it meets an obstacle in its path, which possesses a high positive charge concentrated at a single point.

On the basis of his own research work, Rutherford in 1911 proposed a new "planetary" model which compared the atom to the solar system. At the center was a small positively charged 'nucleus' containing almost the whole mass of the atom. The nucleus was surrounded with electrons, the number of which was determined by the magnitude of the positive nuclear charge. However, such a system can be stable only if the electrons are moving since otherwise they would fall upon the nucleus. Consequently the electrons of the atom must revolve about the nucleus in approximately the same way as the planets move around the sun.

Diagram of the Rutherford's experiment is shown in Fig. IV. A narrow beam of alpha particles was directed at a leaf of thin metal foil M. Their further behavior could be observed by means of a device R for registering alpha particles, which moved along the arc A.

The diagram for deflection of alpha particles by a nucleus is given in Fig. V. on next page.

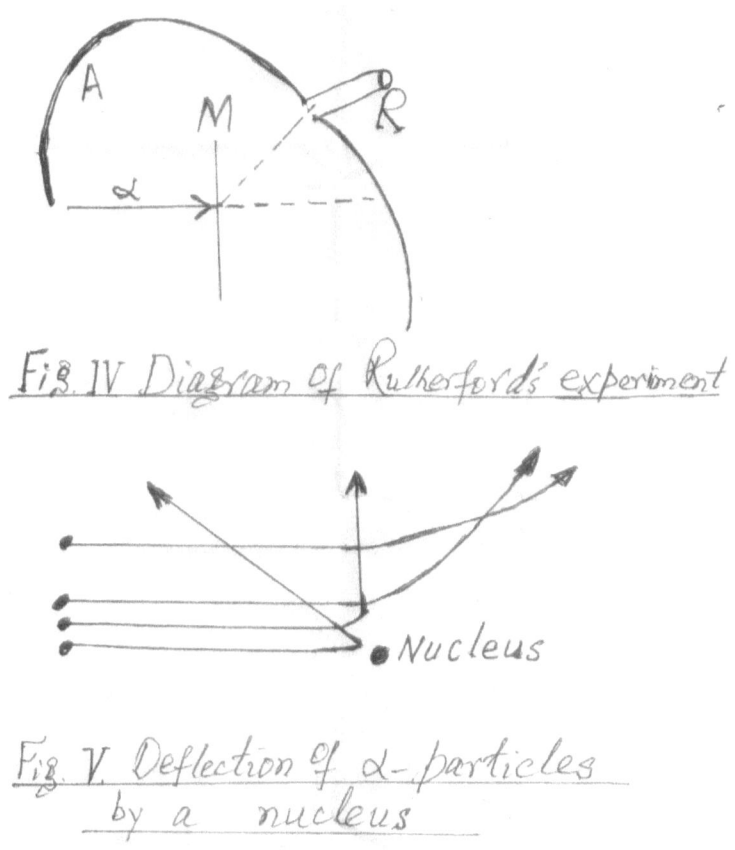

Fig. IV Diagram of Rutherford's experiment

Fig. V Deflection of α-particles by a nucleus

In order to explain large deflection of alpha particles by the atom, Rutherford proposed that the entire positive charge of the atom is concentrated at the center, called the nucleus of the atom.

In 1932, J. Chadwick discovered neutron which is a neutral particle of zero charge and of mass approximately equal to that of proton. Protons and neutrons are collectively referred to as 'nucleons'.

Rutherford was awarded the 1908 Nobel Prize in chemistry for his theory of atomic structure. Physicists had to choose either electrodynamics or the planetary atom. They silently chose the former, because a revolving electron must inevitably and rapidly fall into the nucleus.

Rutherford's experiments could neither be forgotten nor disproved. Physics had run into a blind alley. There had to be Niels Bohr to find the way out.

4. From Classical Mechanics to Quantum Mechanics

Classical mechanics was born in the year 1687 when Newton's book "Philosophia Naturalis Principia Mathematica" ("The mathematical Principles of Natural Philosophy") appeared in London. In his work Newton formulated for the first time three basic principles of classical mechanics, later called Newton's three laws of motion.

Newtonian Mechanics believes in the principle of mechanical determinism. By the end of the 19th century Newtonian mechanics was in crisis. This crisis signified the fall of universal determinism. Sinister cracks appeared and finally the entire structure began to crumble under the bombardment of new facts.

One of the most fundamental facts was the remarkable constancy of the velocity of light. Another stumbling block to classical physics was the thermal radiation of heated bodies, and then finally the discovery of radioactivity.

Out of these cracks in the structure of classical mechanics grew the theory of relativity and the quantum theory.

English physicists Raleigh and Jeans obtained the unified law of thermal radiation of heated bodies. This law stated that the intensity of radiation emitted by a hot body is directly proportional to the absolute temperature and inversely proportional to the square of the wavelength of the emitted light.

This law was found to be good only for the long-wave portion of the visible spectrum, the green, yellow and red. The law broke down as the blue, violet and ultraviolet rays were approached. This curious situation became known as the "Ultraviolet Catastrophe".

History once again demonstrated that great necessity gives birth to great men. The way out of this cul-de-sac of classical physics with its immutable dogmas was found by Max Planck who in 1900 introduced the concept of quanta and by Albert Einstein, who in 1905 advanced the theory of relativity.

Max Planck, a German Physicist discovered the Quantum Theory that revolutionized our understanding of atomic and subatomic processes, Just as Albert Einstein, again a German physicist developed the Theory of Relativity that revolutionized our understanding of space and time.

Max Planck in 1900 explained the distribution of energy in the spectra of heated bodies, by developing a theory based on the assumption that energy is not emitted by atoms continuously but only in minimum indivisible packets called <u>quanta</u>, the value of which depends on the frequency of light emitted.

According to Planck's formula

$$E = h\nu$$

Where E = energy of quantum

ν = frequency of vibrations (V, pronounced as Nu).

h = Planck's Constant = 6.626×10^{-34} joule-second

Thus the energy of a body can change by values that are multiples of hν, just as the electric charge can change by a value that is a multiple of the charge of an electron.

The value of Planck's constant is just as great to physics as its magnitude is small. It is this insignificant magnitude of the quantum that makes the light of a candle or the sun appear to us to burn with a constant glow.

Experimental data splendidly confirmed Planck's theory. Planck's formula expresses one of the most important laws of nature. Planck's constant, like the velocity of light and the charge of an electron, is a fundamental constant which cannot be expressed by any other simpler parameter. Planck's discovery of energy quanta won him the Nobel Prize in physics in 1918.

In 1905, Albert Einstein published his theory of <u>photoelectric effect</u> in metals. He explained this effect very simply if light were regarded as a stream of particles called photons. The photons on colliding with the electrons transmit to them their energy equal to hν in accordance with Planck's formula. This also explained why radiation of long wavelengths does not produce the photoelectric effect; the reason for this is that the energy of the photon in this case is not sufficient to tear the electron from the metal.

Those electrons that escaped without giving up to the atoms of the metal any of the energy they received from the photons will have the maximum energy. Obviously, the energy of such electrons is equal to the difference between the energy of a photon hν and the work required to overcome the force retaining the electron in the metal, i.e. the work function of the electron W; hence

$$(E_e)_{max} = h\nu - W$$

This equation, called Einstein's law for the photoelectric effect, is fully consistent with experimental data. Albert Einstein was awarded Nobel Prize in 1921 for his discovery of the law of photoelectric effect.

5. Bohr's Model

On the basis of conventional mechanics and electrodynamics, the planetary system of electrons revolving around the nucleus as suggested by Rutherford could not exist for more than an extremely short period of time. But atoms do exist permanently and do not show any tendency to collapse.

Defying the well-established laws of classical mechanics and electrodynamics, Niels Bohr in 1913 stated that in the case of the motion of electrons within an atom, the following postulates are valid:

1. From all the mechanically possible circular and elliptical orbits of electrons moving around the nucleus ony a few quantized orbits are permitted whose angular momentum is given by:

$$mvr = nh/2\pi \qquad (1)$$

   Where m is the mass of the electron, v is its velocity, r is the radius of the orbit in which the electron is rotating, n is any integer and can take values 1,2,3..........n, h is Planck's constant. That is to say that the angular momentum of the orbit is quantized and is restricted to certain values which are integral multiples e.g. 1, 2, 3.........n of a definite quantity

$$h/2\pi \qquad (2)$$

2. Circulating around these orbits around the nucleus, the electrons are prohibited from emitting any electromagnetic waves, even though conventional electrodynamics says that they should. Such orbits would be consequently stable and would represent stationary states of the atom.

3. Electrons may jump from one orbit to another, in which case the energy difference between the two states of motion is emitted in the form of a single light quantum whose frequency is given by the following quantum theory equation:

$$v = E/h \quad (2) \text{ or } hv = E \quad (3)$$

Where E is the energy difference between the states and v is the frequency of emitted radiation.

If E_2 is the energy of the atom in a state of higher energy and E_1 is the value in the state of lower energy, then an electronic transition from the former to the latter state will be accompanied by the emission of energy E_2-E_1. By Planck's quantum theory, the frequency of the emitted radiation would be:

$$v = \frac{E_2 - E_1}{h} \quad (4)$$

and

$$\lambda = c/v = ch/E_2 - E_1 \quad (5)$$

Where

λ is the wavelength and c is the velocity of light. Each particular energy transition, from ne state to another, should thus result in the emission of radiation with definite frequency given by equation (4) which results in the formation of a spectral line in the spectrum.

The lowest energy state of the atom is called the ground state. The higher energy sates are called excited states. Electrons in the lower energy state can pass to the higher energy state if energy is supplied to them by the use of high temperature or suitable electric discharge. Electrons spontaneously return from higher to lower energy states resulting in the emission of spectral lines.

Bohr Orbits of hydrogen atom and various series of spectral lines are shown in Fig. VI. On next page:

Niels Henrik David Bohr, the Danish physicist who made foundational contributions to understanding atomic structure and quantum theory was awarded Nobel Prize in physics in 1922.

Arnold Johannes Wilhelm Sommerfeld (1868–1951), an outstanding physicist and brilliant teacher was one of the first in Europe who not only accepted Bohr's postulates at once, but developed them further. He reasoned as follows: If an atom is similar to the solar system, then an electron in such a system can travel both along a circle, as in Bohr's model and along ellipses as well, the nucleus being at one of the foci of the ellipses.

Although the Bohr-Sommerfeld theory does explain many characteristics of the spectra, it has many faults as enumerated below:

1. It is based on quantization rules which do not stem from the laws of mechanics and electrodynamics.
2. The use of this theory in calculating a number of spectral characteristics, in particular the intensity of spectral lines and their multiplet structure, yields results that do not agree with those obtained experimentally.
3. The Bohr-Sommerfeld theory when used for the calculation of the energy of electrons in many-electron toms, also fails to give results that coincide with those obtained in experiments (even for the simplest case, the He atom).
4. It was found that the theory could not be used for the quantitative explanation of chemical bonding.

CHAPTER 5

1. De Broglie Waves

The corpuscular properties of light are most clearly manifested in two phenomena: The *photoelectric effect* and the Compton effect.

Einstein showed that the mass of a body m is related to its energy E according to the equation

$$E = mc^2 \qquad (1)$$

Where, c is the velocity of light.

Whereas the photoelectric effect and the Compton effect are clearly indicative of the particle nature of light, the X-ray diffraction, interference and diffraction phenomena are evidence of their wave nature.

De Broglie supposed that not only a light ray, but all bodies in nature must possess both wave and particle properties simultaneously. Therefore, besides light waves and particles of matter, quanta of light and waves of matter must also really exist in nature. De Broglie waves are the foundation of present day quantum mechanics.

De Broglie introduced the idea of 'particle waves' as a consequence of the quantum theory. He pointed out that for photons there are two fundamental equations to be obeyed:

Planck's formula: $E = h\nu$ (1)

Where, E is energy, h is Planck's constant, and ν is the frequency of vibrations.

Einstein Equation: $E = mc^2$ (2)

Where, E is the energy, m is the mass and c is the velocity of light.

Combining equations (1) and (2) and remembering that

$$\lambda = c/\nu \qquad (3)$$

(Where, λ is the wavelength, c the velocity of light and ν the frequency), it follows that

$$\lambda = h/mc \qquad (4)$$

De Broglie then put forward the suggestion that the motion of particles such as electrons was associated with a wavelength given by an expression similar to that for photons viz;

$$\lambda = h/mv \qquad (5)$$

Where, m is the mass of the particle (electron) and v is its velocity.

Equation (5) is the De Broglie equation.

These ideas on "electron waves" received experimental confirmation when Davisson and Germer (1927) and Thomson and Reid (1928), independently showed that a beam of electrons could be made to undergo diffraction and interference effects—these phenomena being interpretable only by attributing wave properties to the electron beam.

Moreover, the observed wavelengths were exactly those required by the De Broglie equation (5). Consequently, not only do light waves behave as streams of small particles (photons), but streams of small particles such as electrons behave as waves. The apparent paradox was resolved by the Heisenberg Uncertainty Principle. De Broglie was awarded Nobel Prize for physics in 1929 for his discovery of the wave nature of electron.

2. Heisenberg's Uncertainty Principle

The Heisenberg uncertainty principle states: "It is *impossible to determine precisely and simultaneously the position and momentum of an electron.*"

This statement can be illustrated by the following discussion. In any measurement of the position of an electron, the radiation used to observe the electron must undergo a change. The measured quantity is the change that occurs in the photon upon contacting the electron. Hence, it is impossible to detect accurately an object smaller than the wavelength of radiation employed to measure it.

As a result, detection of a particle as small as an electron would require very low wavelength, λ, and thus very high energy radiation, $E = hc/\lambda$. However, since the electro is so small, collision with the high energy photon can change the momentum of the electron. As a result, the more accurately the position is measured (by using small wavelength, high energy photons), the less accurately can the momentum be measured simultaneously and vice versa.

If Δx is the uncertainty in defining the position and Δv the uncertainty in the velocity, the uncertainty principle may be expressed mathematically

$$\Delta x \cdot \Delta v \geq h/4\pi$$

Where, h = Planck's constant.

The Bohr model violates the uncertainty principle, for it describes simultaneously both the location and momentum of the electron. This concept must therefore, be replaced by the probability of finding an electron in a particular position, or in a particular volume of space.

Werner Heisenberg was awarded Nobel Prize in physics in 1932.

3. Schrodinger's Wave Equation

The founding fathers of Quantum Physics derived inspiration from Vedas. While formulating their groundbreaking theories, they sumptuously dug into annals of Vedic philosophy and found their experiments to be consistent with the knowledge expounded in Vedas.

Danish physicist and Nobel laureate Niels Bohr was fascinated with Vedas. His remark, "I go to the Upanishad to ask questions", reveals a lot about his respect for the ancient wisdom of India.

Erwin Schrodinger, an Austrian-Irish physicist who won the Nobel Prize for his famous wave equation was also a keen proponent of the Vedic thought.

After the De Broglie proposal, Schrodinger intuitively selected an equation for a wave as the model to describe the behavior of an electron in an atom. This model incorporates the requirements stated in the uncertainty principle.

To review the mathematical description of wave motion, consider the wave illustrated in Fig. IA

This wave, after a time interval, progresses along the x axis in the direction indicated by the arrow to the position illustrated in Fig. IB This wave motion can be described quantitatively by the differential equation:

$$\delta^2 A / \delta x^2 = (1/c^2)(\delta^2 A / \delta t^2) \qquad (1)$$

Where,

A is the amplitude, (i. e; the height of the wave measured (along the y axis) at a particular distance x along the x axis; c is the velocity at which the wave is travelling, and t is the time.

The differential equation (1) has a solution:

$$A = a \sin 2\pi(x/\lambda - vt) \tag{2}$$

Where, λ is the wavelength, v is the frequency, and a is a constant. The amplitude A, for a time t and position x, can be calculated from (2).

Another type of wave equation, a standing wave, results when a string whose ends are fixed is plucked. A standing wave results in a stationary pattern (see Fig. II) where the actual profile is fixed instead of travelling along the x axis as the wave illustrated in Fig. 1B does. Empirically, it is found that this model (extended to a three-dimensional wave as in Equation (4) best describes the behavior of an electron attracted by a nucleus.

The differential equation used to describe this one-dimensional wave is similar to equation (1) and has the solution (3)

$$A = f(x) \cos 2\pi vt \tag{3}$$

Where,

f(x) is an abbreviation for $2a \sin 2\pi x/\lambda$ and is a function of the x coordinate only.

The second derivative of equation (3) with respect to time can be obtained and combined with equation (1) to eliminate the quantity t. the resulting equation for the standing wave can be converted to equation (4) to describe a three-dimensional wave:

$$\delta^2 \Psi/\delta x^2 + \delta^2 \psi/\delta y^2 + \delta^2 \psi/\delta z^2 + 4\pi^2 \psi/\lambda^2 = 0 \tag{4}$$

Where, ψ is the counterpart in the three-dimensional system of A in equation (1).

Abbreviating

$$\delta^2/\delta x^2 + \delta^2/\delta y^2 + \delta^2/\delta z^2$$

As ∇^2, one obtains:

(Where, ∇ is called Laplacian Operator),

$$\nabla^2 \psi + 4\pi^2 \psi/\lambda^2 = 0 \tag{5}$$

Equation (4) or (5) is a differential equation similar to (1) except that this equation specifically relates to a stationary, three- dimensional wave and has the variable t eliminated from it. This variable is eliminated to produce an equation whose solutions are time independent. The differential equation (4) can be solved more easily than those that contain t.

Schrodinger selected the mathematical description of a standing wave as the basis of his model for the structure of an atom. He incorporated into the expression for a standing wave (equation 5) the de Broglie assumption, $\lambda = h/mv$, and obtained:

$$\nabla^2 \psi + (4\pi^2 m^2 v^2 /h^2) \psi = 0 \qquad (6)$$

By combining equation (6) and the equation relating total energy, E, potential energy, V, and kinetic energy, $½ mv^2$ i.e;

$$E = V + ½ mv^2 \text{ or } v^2 = 2(E-V)/m \qquad (7)$$

One obtains an equation:

$$\nabla^2 \psi + (8\pi^2 m/h^2)(E-V) \psi = 0 \qquad (8)$$

Equation (8) is the <u>Schrodinger Wave Equation</u>.

It is well to remember that equation (8) is not derived from first principles but is the equation which results from:

(a) Empirically selecting the equation for a standing wave as the model for the behavior of an electron in an atom and

(b) Incorporating the de Broglie assumption.

Justification for the above "derivation" is found in the fact that the solution of equation (8) yields values for the energy, E, which correspond closely to those obtained experimentally from atomic spectra.

The symbol ψ in equation (8) is of great significance. Since ψ is analogous in three dimensions to A (the amplitude of a planar wave), ψ is referred to as the amplitude function. Physical significance cannot be attributed to ψ but $\psi \psi^*$ may be shown to be proportional to the probability of locating the electron at a given position (here ψ^* is the complex conjugate of ψ). The quantity $\psi \psi^* d\tau$ gives the probability of finding the electron in a volume element $d\tau$. If ψ is a real number, $\psi \psi^*$ becomes ψ^2.

The solutions which satisfy the following requirements are the only meaningful solutions in terms of describing the behavior of an electron in an atom:

(a) The wave function must be finite and continuous, i. e; it can never approach infinity for any value of parameters to locate a point in the spherical coordinate system relative to Cartesian Coordinates.

(b) The solution must be single valued. i. e; at a given point there can never be more than one value for the amplitude, ψ,

(c) The solutions must be normalized. This requires that if a solution, ψ, is squared, multiplied by d τ, and integrated over all space, the result must equal 1; i. e;

$$\int_0^{+\infty} \psi^2 \, d\tau = 1 \tag{9}$$

Equation (9) indicates that Ψ must approach zero as τ becomes infinite.

Since Ψ^2 d τ is related to the probability of finding the electron in volume element, dτ, the integration in equation (9) simply requires that there be unit probability of finding the electron somewhere in space. There are only a limited number of solutions to the Schrodinger equation for an un-ionized atom that satisfy all of the above requirements. The allowed solutions are called <u>eigenfunctions</u> and each one represents an <u>orbital</u> with a capacity of two electrons in the atom.

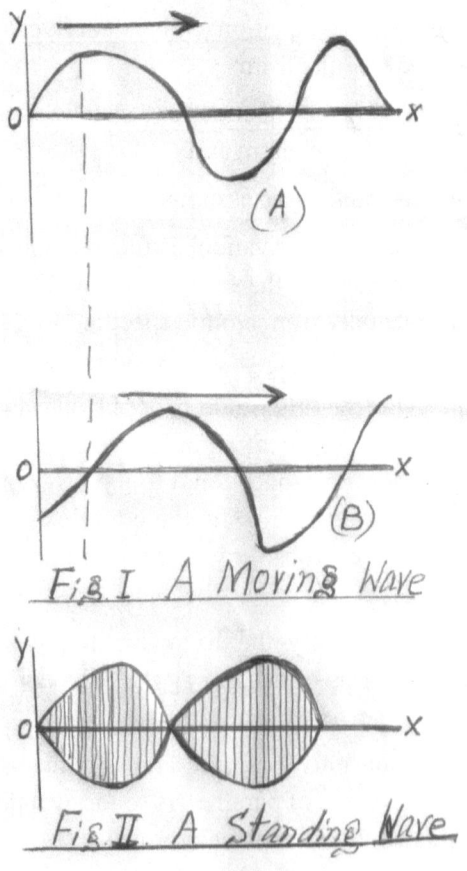

Fig. I A Moving Wave

Fig. II A Standing Wave

4. Quantum Numbers

The quantum –mechanical treatment of various cases of motion of micro particles within a 'confined region of space' (say, in an atom, molecule etc.) shows that *the wave function of a particle always contains dimensionless parameters which can take a number of integral values*. These values are called quantum numbers.

The expression 'confined region of space' implies that the particle is confined to a given region by the action of some forces and the probability of finding it beyond this region is close to zero.

The number of quantum numbers contained in the solution is equal to that of the *degrees of freedom* of the particle. The *number of degrees of freedom is the number of independent components of motion of the particle*. Thus, in one-dimensional potential square well, the particle has only one degree of freedom; in the case of translator motion in space, it has three degrees of freedom (motion is possible in the direction of each of the three coordinates x, y and z); if besides this, the particle can rotate round its own axis, it has fourth degree of freedom.

Acceptable solutions to Schrodinger wave equation (8), that is solutions which are physically possible, must have the following properties:

1. ψ must be continuous.
2. Ψ must be finite.
3. Ψ must be single- valued.
4. The probability of finding the electron over all the space from plus infinity to minus infinity must be equal to one.

The probability of finding an electron at a point, x, y, z is Ψ^2, so

$$\int_{-\infty}^{+\infty} \Psi^2 \, dx \, dy \, dz = 1$$

Several wave functions called Ψ_1, Ψ_2, Ψ_3......... will satisfy these conditions to the wave equation, and each of these has a corresponding energy E_1, E_2, E_3.....

Each of these wave functions Ψ_1, Ψ_2 etc. is called an **orbital**, by analogy with the orbits in the Bohr Theory. In a hydrogen atom, the single electron normally occupies the lowest of the energy level E_1, which is called the ground state. The corresponding wave function $\Psi 1$ describes the orbital that is <u>the volume in space where there is a high probability of finding the electron</u>.

The four quantum numbers are:

1. Principal Quantum number (n)
2. Azimuthal Quantum Number (l)
3. Magnetic Quantum Number (m)
4. Spin Quantum Number (s)

The numerical values of n, l, m, and s, are given below;

$n = 1, 2, 3\ldots\ldots\ldots\ldots\ldots n$

$l = 0$ to $(n-1)$

$m = +1, 0, -1$

$s = \pm\frac{1}{2}$

Each value of m has two values of s, i.e; $+1/2$ and $-1/2$, and each value of s represents one electron.

When $l = 0$, the orbital is spherical and is called a s-orbital;

When $l = 1$, the orbital is dumb-bell shaped, and is called a p-orbital.

When $l = 2$, the orbital is double dumb-bell shaped and is called a d-orbital.

When $l = 3$, a more complicated f-orbital is formed.

The letters s, p, d, and f come from the spectroscopic terms sharp, principal, diffuse and fundamental, which are used to describe the lines in the atomic spectra. The various atomic orbitals with the number of electrons are given in the Table I below:

n l m symbol

1 0 0 1s (one orbital)

2 0 0 2s (one orbital)

2 1 –1, 0, +1 2p (three orbitals)

3 0 0 3s (one orbital)

3 1 -1, 0, +1 3p (three orbitals)

3 2 –2, –1, 0, +1, +2 3d (five orbitals)

4 0 0 4s (one orbital)

4 1 –1, 0, +1 4p (three orbitals)

4 2 –2, –1, 0, +1, +2 4d (five orbitals)

4 3 –3, –2, –1, 0, +1, +2, +3 4f (seven orbitals)

Shapes of orbitals are given in Fig. I, II and III.

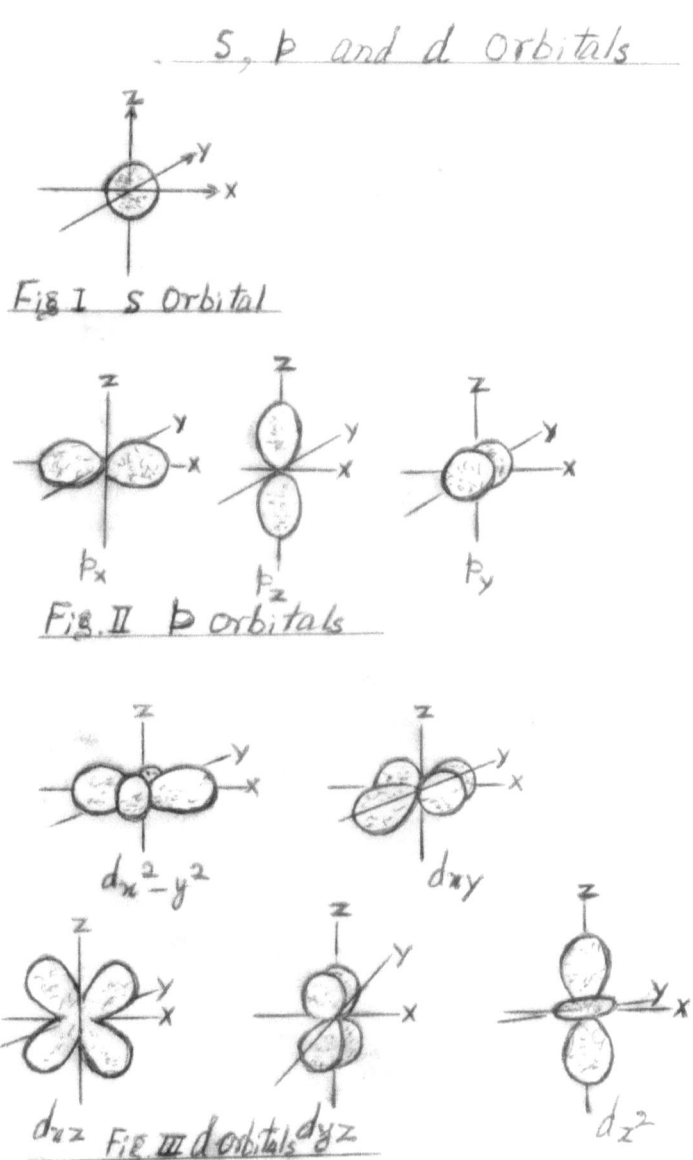

S, p and d Orbitals

Fig I s Orbital

Fig. II p orbitals

Fig. III d orbitals

The three 2p orbitals are all of equal energy. The term "triply degenerate" is employed to describe the equivalence in energy of the three orbitals. Each orbital can accommodate two electrons and these are distinguished by the spin quantum number, s of values $+1/2$ and $-½$.

The relative energies of the orbitals vary with change in nuclear charge of the atoms. The following order can be employed for writing the atomic structure of most of the atoms in the periodic table:

$$1s<2s<2p<3s<3p<4s<3d<4p<5s<4d<5p<6s<4f<5d<6p<7s$$

The order pertains to the hypothetical synthesis of an atom by simultaneous addition of a proton and an electron to the preceding element. This order is known as <u>Aufbau Principle</u>. "Aufbau" literally means "to build up".

As the main quantum number increases, the difference in energy between the orbitals decreases. Beyond 7s there are only very slight differences. However for transition elements the energy sequence obtained by the Aufbau Principle does not apply.

A more nearly complete description of the electrons in an atom can be obtained by employing all four quantum numbers, n, l, m and s. A box diagram representation is one method employed to indicate all four quantum numbers. Usually only the valence electrons are indicated. The s orbitals are indicated by a single box ☐. The p orbitals are indicated by three boxes ☐☐☐, one each for p_x, p_y and p_z orbitals. The d orbitals are represented by five boxes, ☐☐☐☐☐, one each for $d_{x^2-y^2}$, d_{xz}, d_z^2 d_{yz}, d_{xy}, and the f orbitals by seven boxes.

A spin quantum number of $+1/2$ is indicated by the symbol 'half arrow' pointing upward and $-1/2$ by 'half arrow' pointing downward.

In addition to the above conventions, two rules are required for the application of the box diagram designation to all atoms:

(1) <u>Pauli exclusion principle:</u>

This principle states that no two electrons in an atom can have all the four quantum numbers same.

(2) <u>Hund's rule:</u>

This rule states that electrons will distribute themselves in degenerate orbitals so as to retain parallel spins as much as possible; i.e; electrons do not pair up until they have to.

References

1. M.M.WILLIAMS, INDIAN WISDOM, INDIGO BOOKS.
2. M. KARAPETYANTS, S. DRAKIN, THE STRUCTURE OF MATTER, MIR PUBLISHERS MOSCOW
3. A.K.BARNARD, Theoretical Basis of Inorganic Chemistry, McGraw-Hill Publishing Company.
4. J.D.LEE, a new CNCISE INORGANIC CHEMISTRY, English Language Book Society and Van Nostrand Reinhold Company Ltd. LONDON.
5. RUSSEL S.DRAGO, Physical Methods in INORGANIC CHEMISTRY, EAST- WEST PRESS
6. V.J. RYDNIK ABC's of Quantum Mechanics.
7. JAMES E. HUHEEY, INORGANIC CHEMISTRY, Harper International SI Edition
8. Cotton and Wilkinson, Basic Inorganic Chemistry, Wiley India.

PART III

CHEMISTRY TEACHING AND LEARNING

Our educational institutions in general and those concerned with higher education in particular have over the years come in for much criticism and public denunciation. The educational structure, by and large, has continued unchanged since the days of its inception when the circumstances of British rule created a rigid and authoritarian system which automatically precluded any infusion of dynamism or questioning content.

An obsolete and irrelevant curriculum has successfully stunted the mental growth of students. Indian students are generally apathetic about their work. There is little intellectual curiosity or desire for knowledge. A degree is, in the final analysis, only a document of release. For boys it is: degree or job, whichever is earlier. For girls it is: degree or marriage, whichever is earlier.

It is escapist to say that this malaise is caused by the students themselves. To find the root cause it is necessary to look beyond: at the limited syllabi, un-imaginative text books and archaic methods. This malaise has of course been cured to some extent in this digital age, but still the internet remains out of access most of time especially in far flung areas.

An understanding of syllabus that encourages learning by rote lends an air of lethargy to most of our educational institutions. Mental faculties are blunted by the unquestioning attitude instilled in students. The imagination or interest of a pupil is seldom captured and his curiosity is never given a fillip.

The biggest nuisance still prevalent in our educational institutions is the malpractice of "mass copying". This curse happens in broad daylight in collaboration of the men whom we call "Builders of the Nation", the teachers. There are some instances of students having been found to manage to top the Board or University examinations. The hand of a higher-up or an executive or even a politician has been found to be involved in such cases.

The syllabus presents a considerable hurdle which becomes insurmountable in the face of the present examination system. Students are questioned on a water-tight portion that is never stretched. The under graduate is not encouraged to study new interpretations or analyses as time has stood still for most universities.

How long will Indian Colleges remain in stagnant backwaters?

Our educational system is like an open system, to be more precise—a four-component system. Student, Curriculum, Society, and Teacher constitute its four components **(SCST System)**

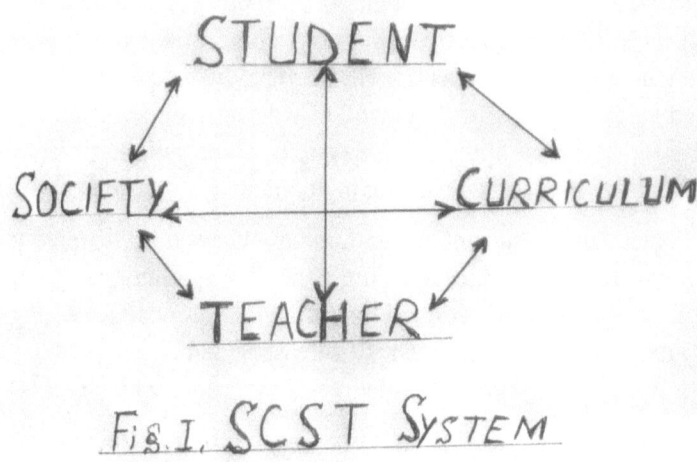

Fig. 1. SCST System

Whatever is pertinent, or opposite, to anyone factor inevitably affects all others. The teacher mediates the curriculum to the student, the society produces and to some extent conditions them and it employs the teachers. The students will become that society and will bring the cultural and technical insights that they have learned. What is relevant to one part affects the others.

Yes, the curriculum and the courses of study have evolved with time but the more things change, the more they remain the same. A large chunk of students attend schools and colleges for the sake of attendance. Their attendance in educational institutions is a mere time-pass, since they are laden with ready-made Xerox copies of the answers from private teaching shops, purchased on a regular basis at fees well within the range of their loving but least caring parents.

The irony of fate is that in every school and college there is a fair percentage of teachers who shirk work. They are conspicuous by their lack of devotion and dedication and continue to draw fat salaries from the state exchequer. This graphic when projected on national level becomes catastrophic. The politicians on their part are also responsible for this "educational Catastrophe".

It is axiomatic that no one today can be considered literate without a sound knowledge of what science is, what it has contributed to our lives, and how it affects our technologies (scientific and humanistic). This knowledge should be based on some substantive scientific experience. Much of our present education alienates our students from their culture. There is increasing support for cults of irrationality and nihilism. Our present approaches to science education are not working as we would like them to.

The moral education in our schools and colleges education is conspicuous by its absence. The atmosphere in our educational institutions is a sorry state of affairs. Besides the malaise of mass copying, there have been instances of rape in the temples of knowledge. Our head hangs in shame on hearing these heinous crimes occurring in broad daylight in the presence of our Nation Builders themselves. The height of all is that in some cases the school teachers and principals are also involved.

Turning now to specific Chemistry curricula, we find ourselves faced with a dilemma with which countless committees have struggled for years: when we stick to broad generalities we can agree, but as we attempt more detailed specification of curricular content the area of agreement constricts. Because so much of our previous effort in chemical education has been focused on content, we have tended to neglect the problems of communication and the personal problems of students in learning the meaning of science and sensing its power to strengthen their lives.

An important observation is that the amount of substantive material available in science is fantastically large—literally beyond individual comprehension. The school bags are too heavy for the student. There is a large number of drop-outs in-spite of the attempts at universal free primary and secondary education. The emphasis on behavioral objectives in curricular development is very important. The heavy load of home-work is another impediment in the all- round development of student personality. In a class of fifty or more students the individual attention is a far cry in the wilderness. The condition of laboratories is mostly pathetic in our schools and colleges.

Chemistry has been defined as the science of atoms in action, not of atoms in specimen bottles on the shelves of a museum. One consequence apparent to teachers is that courses evolve not only from a course content point of view, but also in method and technique. Furthermore, teachers change their minds about critical topics at least as often as new trends are developed by the society and its student populations.

A pertinent example of changing times is the case of energy. Energy consumption and the threat of an energy shortage have brought basic chemical factors to the attention of most people. If as chemists and educators, we are to take advantage of this, we must begin to consider these factors in our classes.

Since ancient times man has endeavored to exploit environmental energy. The environmental energy includes: 1. Solar energy. 2. Geothermal energy. 3. Chemical energy of fossil fuels.

The present sources of energy are not ever-lasting. They will be considerably exhausted in not too distant future.

Now if our earth is to support the population explosion accompanied by some increase in living standard, it becomes inevitable to find new sources of energy.

A student of chemistry knows that the nucleus of an atom contains protons and neutrons, collectively called nucleons. According to Coulomb's law the positively charged protons in the nucleus must repel each other with very large force because of their nearness and the nucleus must break into pieces. But this does not happen.

It means that there is some other force in the nucleus which overcomes the electrical force of repulsion between the protons and binds the nucleons inside the nucleus. This force is called <u>nuclear force</u> and is the strongest force so far discovered.

Einstein's equation $E = mc^2$ tells us that mass can be converted into energy and vice versa. The total energy required to tear apart a nucleus into its constitutional protons and neutrons is called the nuclear binding energy.

Binding energy curve is shown in Fig. I:

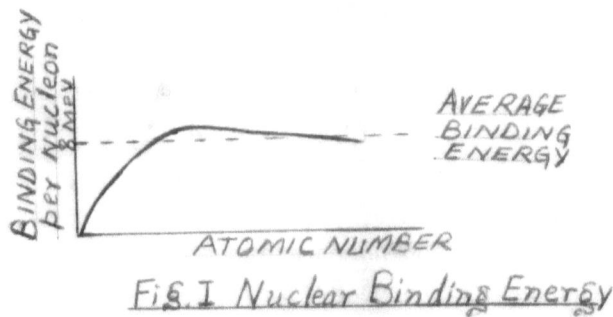

Fig. I Nuclear Binding Energy

The Binding Energy (B.E.) of the nucleus varies with the number of nucleons present in it, and to compare the nuclear stability of different elements the average binding energy per nucleon is calculated.

B.E. per nucleon = Total B.E./Number of nucleons

The B.E. for each additional nucleon decreases as the nuclei get larger.

The fact that binding energy curve "droops" at both high and low mass numbers has practical consequences of the greatest importance. The curve of binding energy has even been adopted as the title of a book (by John McPhee) about the possibilities of Nuclear Terrorism.

The drooping of the binding energy curve at high mass numbers tells us that energy can be released if a single massive nucleus ruptures into two small fragments. This process is called Nuclear Fission.

The drooping of the binding energy curve at low mass numbers, on the other hand tells us that energy will be released if two nuclides of small mass number combine to form a single middle-mass nuclide. This process, the reverse of fission, is called Nuclear Fusion.

The discovery of nuclear fission in 1939 saw the advent of atomic age. It revealed a new and highly concentrated source of energy. You must be clear whether your concern is for energy or for the rate at which the energy is delivered, that is for power.

In the nuclear case will you burn your kg of uranium slowly in a reactor or explosively in a bomb? A series of publications brought out by the International Atomic Energy Agency (IAEA) from time to time give good account of the applications of nuclear science:

I. Isotopes as Tracers in Agriculture and Biology.

II. Nuclear Irradiaton.

III. Nuclear Analytical Techniques.

IV. Nuclear Power.

The Fourth Biennial Conference On Chemical Education (the University of Wisconsin—Madison) summarized the aims and purposes of chemical education as follows:

1. All colleges are seeking answers to bothersome problems.
2. Students should become aware of what chemists do, and how they do it.
3. Students should gain some skill at the practice of chemistry since it is a laboratory-oriented discipline.
4. There is some measure of concern for ethical considerations (values) as related to the practice and teaching of chemistry.
5. There should be a component of 'real chemistry' in most courses. Descriptive chemistry, particularly as it relates to industrial chemistry offers an excellent vehicle to accomplish this.

6. Chemistry students must be helped to develop some critical skills in communication (reading, writing, speaking) and especially areas like electronics and computers.
7. Teachers should periodically review their course objectives and expectations to keep them current.
8. No course is identical to one presented in another place, at another time—nor should we expect it to be. Courses should continually evolve.

The first two years of a chemistry course are critically important. Environmental problems are seen as a useful framework in which to teach traditional chemistry especially at the introductory level. Particularly in the first year, special efforts need to be made to give students an appreciation for, and historical insight into the immense impact of chemical science on thought and technology. The impact of chemical science in agriculture, medicine, industry and other segments of modern technological society should get due emphasis.

Lecture experiments should be made easier and simpler. Chalk is used in every classroom. When you use it you are experimenting with chemical bonding. Bonds break and form on writing with calcium carbonate $CaCO_3$ on the blackboard, when you erase and when you wash the board. Students preparing for professional work in chemistry need a good grounding in integral and differential calculus.

The quality of an undergraduate education is more important than its precise content. However, a successful career in chemistry requires both breadth and opportunity for specialization. An essential ingredient in addition to quality is time. To become a chemist takes extended and diligent efforts.

Laboratory work requires gaining experience with a variety of modern techniques in synthesis and analysis, structure identification and determination, chemical kinetics etc. A student should achieve self-confidence in the ability to perform quantitative manipulations and with understanding a good selection of modern instruments. He should plan experiments through use of literature, observe standard safety measures, keep adequate experimental records and write good reports.

Lecture halls for chemistry courses and staff rooms, near staff and student research laboratories should be provided in buildings equipped with modern demonstration facilities.

Laboratories should be well ventilated and equipped with necessary facilities like gas, water and electric power. Hoods should be readily available.

Library should have rich collection of text books, reference books, periodicals and current journals with good back runs.

The instruments and equipment required for modern undergraduate education in chemistry include typically the following: Single pan analytical balance, pH meters, recording spectrophotometers, gas chromatographs, colorimetry, conductance and automatic temperature control and a polarography.

Safety in a chemistry laboratory must be given the top priority. Instruction in chemical safety, maintenance of modern safety devices and observation of standard safety practices in the laboratory should be emphasized. Special facilities are needed for handling, storage and disposal of hazardous chemicals.

The training of chemistry teachers is an essential part of chemical education. The following points highlight this important part of education:

1. Each university chemistry department should accept the responsibility for the orientation of college teachers. Administration should take an active role in implementing interaction between higher 10+2 school and college chemistry teachers. College chemistry departments should work cooperatively with local secondary school teachers in planning and conducting all-day seminars.
2. Broadest possible encouragement should be given for the development of innovative experimental programs leading to better trained teachers and for the careful study and evaluation of the results of these programs.
3. Further academic training is highly desirable. It should stress upon both depth and breadth but need not particularly stress research.
4. A very sound background in all areas of chemistry and its relationship to other sciences and to the society must be strived.

The method of assessment is at the core of any education. It is not sufficient to check student's knowledge at the end of an academic year with archaic questions and ready-made answers. This system has led to extremely selective studying by the students.

An episodic knowledge of a subject is not adequate, where the concepts, ideas and trends form only a hazy background. It has given us a set of students who are lethargic and a faculty that is inert.

Education is not the acquiring of bookish knowledge alone; more important, it is the drawing out of abilities and nurturing of talent.

The emphasis must be shifted from teaching to learning. A teacher should himself be always a learner in view of new and fast developments and advances in science and technology.

One fool-proof method of assessing a student's merit is through his cumulative record which clearly indicates where his abilities, aptitudes, interests and intellectual capabilities lie. Fortunately, the old system of evaluation is yielding place to the modern and fool-proof system of assessment.

Yazachew Alemu Tenaw has beautifully elucidated the importance of Information and Communication technology in chemistry teaching and learning in his article in International Journal of Education Research and Reviews (Vol. 3(3), May 2015 :

The Information and Communication Technology (ICT) opens up a new educational world of creativity for students and teachers. ICT plays an important role in planning lessons and in their management (Grimaldi and Rapauno, 2009).

The use of ICT could be divided into two groups: in the first group a computer is used as a tool for finding information, communication and multimedia and in the second group the computer is a scientific tool such as a virtual laboratory, interactive simulation, computer-assisted laboratory work (Sorgo et al; 2007).

The use of computers in science subjects, particularly chemistry has some specific advantages. Cognitive psychologists assume that the understanding of chemistry includes the ability to think on three levels: The macroscopic level, the symbolic level and the level of particles (Johnstone, 1991).

Teachers should develop a context for learning. For example, students could work in teams to investigate local air quality, learn the nutritional value of their favorite foods or discover the effects of fertilizer on water quality.

In chemistry, well-planned lessons include effective questions, student interaction with new ideas, and student reflection—all focused on the conceptual learning goal. Chemistry teachers should capitalize on the importance of chemistry in everyday life to engage their students and then follow through with opportunities for them to actively explore newly introduced concepts. Advance planning will reap big payoffs in student motivation and deepen their understanding of topics in chemistry.

In this digital world of today, it is high time that the hopes and aspirations of our youth were realized.

References

1. Journal of Chemical Education, Vol.54, 1–5, 1977(6)
2. Journal of Chemical Education, 49, 1–4,6, 1972, 34
3. Jl. Chem. Educ. 50, 1–6, 14, 32, 1973
4. Jl. Chem. Educ. 50, 1–6, 1973, 23, 33, 35
5. Jl. Higher Educ. 3, 3, 1978
6. Yachew Alemu Tenaw, International journal of Education Research and Reviews, Vol.3 (3), pp 078–084, May 2015

www.ingramcontent.com/pod-product-compliance
Lightning Source LLC
Chambersburg PA
CBHW030755180526
45163CB00003B/1038